D0405589

FINDING FIBONACCI

FINDING FIBONACCI

The Quest to Rediscover the Forgotten Mathematical
Genius Who Changed the World

KEITH DEVLIN

PRINCETON UNIVERSITY PRESS
PRINCETON AND OXFORD

Requests for permission to reproduce material from this work should
be sent to Permissions, Princeton University Press

Published by Princeton University Press, 41 William Street,
Princeton, New Jersey 08540

In the United Kingdom: Princeton University Press, 6 Oxford
Street, Woodstock, Oxfordshire OX20 1TR

press.princeton.edu

Jacket art courtesy of Shutterstock

ISBN 978-0-691-17486-0

Library of Congress Control Number 2016962928

British Library Cataloging-in-Publication Data is available

This book has been composed in Gentium Plus

Printed on acid-free paper ∞

Printed in the United States of America

1 3 5 7 9 10 8 6 4 2

Contents

PRELUDE
Sputnik and Calculus   1

CHAPTER 1
The Flood Plain   5

CHAPTER 2
The Manuscript   18

CHAPTER 3
First Steps   35

CHAPTER 4
The Statue   42

CHAPTER 5
A Walk along the Pisan Riverbank   56

CHAPTER 6
A Very Boring Book?   64

CHAPTER 7
Franci   72

CHAPTER 8
Publishing Fibonacci: From the Cloister to Amazon.com   85

CHAPTER 9
Translation   97

CHAPTER 10
Reading Fibonacci   116

CHAPTER 11
Manuscript Hunting, Part I (Failures)   138

CHAPTER 12
Manuscript Hunting, Part II (Success at Last)   151

CHAPTER 13
The Missing Link   167

CHAPTER 14
This Will Change the World   181

CHAPTER 15
Leonardo and the Birth of Modern Finance   192

CHAPTER 16
Reflections in a Medieval Mirror   213

APPENDIX
Guide to the Chapters of Liber abbaci   228

BIBLIOGRAPHY   236

INDEX   239

FINDING FIBONACCI

Sputnik and Calculus

If my name is familiar to you, chances are it's because you have read one of my mathematics books (I've written around 35, many of them for the general reader), or perhaps you've read an article I have published in a magazine or newspaper, read one of my blogposts (I maintain four blogs), taken part in one of the sessions of my massively open online course (MOOC) on mathematics, or heard me on National Public Radio, where I am known as "the Math Guy." Yes, I am *that* Keith Devlin. And yes, I love math.

By most people's standards, I'm also pretty good at math. But it wasn't always so. In the first few years of elementary school, I was one of the poorest performing children in mathematics. In fact, I was the last kid in the class to memorize my multiplication tables, a rite of passage that loomed large in the mathematics education mandated in 1950s England. Mastering the multiplication tables (today's math educators call them "multiplicative number bonds," presumably to give them a more modern ring) caused me so much anxiety that my parents had to go to the teacher to explain that my abysmal performance was not due to lack of effort on my part. Rather, I simply did not get it. I was not, they said, a math type.

Things improved by the time I was aged 9, and as a result of precocious classroom performances in all disciplines, including math, I skipped year 10 and went straight into year 11, the final year of elementary school, taking the national "11 Plus" selection exam for secondary school placement at age 10. I was the only kid at that school to "pass" that exam, meaning I was allocated a place in an academically focused secondary school (a Grammar School), rather than the more common Secondary Modern School, designed to prepare the next generation of British factory workers, the kind of school to which all (and I mean all) of my classmates went on to attend.

Looking back, my early difficulty with math was likely because I got stuck trying to *understand* multiplication, whereas everyone else simply memorized the tables like a meaningless passage from Jabberwocky. But back then, no one else in my world thought of mathematics as something you understand; rather, it was a meaningless memorization hurdle you had to master to progress through the system, justified with the mantra "You will need it in later life." (In my case, it turned out I actually did make use of all that school math, including the multiplication tables, but that was because I became a mathematician. By the time I was at university, electronic calculators had arrived on the scene, rendering obsolete the one demonstrably practical advantage of mastery of the multiplication tables.)

I entered grammar school in 1958 (by then having reached the official age of 11), the year Russia launched the first satellite (Sputnik) into space. For this science-fiction-mad young male, that event was dramatic. Humankind was about to enter the Space Age, and I wanted to be part of it, the more so when the TV news showed US president John F. Kennedy announce that they (the

Americans, still riding high after saving Europe from the Nazis in the Second World War and giving everyone seductive cinema entertainment) were going to put a man on the moon.

So I began my high school career with the goal of becoming a "space scientist." I wasn't entirely sure what that meant, but physics was clearly a part of it. And since you needed math to do physics, I realized that math was also something in which I had to excel. But whereas physics made sense to me, mathematics did not. I did not particularly like it, viewing it mostly as a disjointed collection of recipes and tricks for solving numerical problems. I put effort into it mainly because it was part of my path to becoming a "space scientist."

Sure, I liked that feeling of satisfaction you get when you solve a math problem, but I did not view mathematics as a single discipline the way I saw physics. Math was, to me, just a useful toolbox for doing physics. But then everything changed when, at age 16, I started to learn calculus.

Calculus is unbelievably powerful. It is the mathematical tool that allows you to formulate and solve the equations required to put a man on the moon. Yet at the same time it is deeply mysterious.

As I learned much later in life, the mystery came from the way it was taught. The math teacher provided us with a list of symbolic rules for calculating derivatives and integrals. Master those rules, and you can solve all the equations of space flight. But those symbolic rules were just, well, weird. Why should the slope of the curve x^n be nx^{n-1}? And it was just too simplistic to believe that the slope of $\sin(x)$ was $\cos(x)$. Not even Hollywood movies would be so crass. Yet, according to my calculus textbook, that was the rule.

Still, being older and wiser than the five-year-old me, the sixteen-year-old me did not let my lack of understanding hold me back, the way it had with multiplication. I just practiced applying all of the rules until I could solve all the problems in the calculus textbook. (In the process, I actually did *solve them all*.)

But I was not content to simply use those rules. I really did want to understand them. In the course of striving for that understanding, I came to see mathematics not as a collection of disjointed techniques, procedures, and tricks, but as a single, coherent whole, a vast, cognitive landscape, built up over three millennia by some of the smartest people the world has ever seen. It was a landscape that was both powerful and rich in beauty.

By the time I graduated from high school at age 18, I no longer wanted to be a space scientist. I had become a mathematician, in spirit if not yet expertise. Compared to the mathematical universe I had glimpsed a few parts of—mostly by devouring the few popular mathematics books then available, a genre I would spend a large part of my adult life writing—"Outer Space" seemed boring.

Which is why I entered university to study not physics, but mathematics. Six years later, I had earned both a bachelor's and doctoral degree in mathematics (along the way finally gaining an understanding of calculus, the goal that had first led me to take that path), and I began my career as a professional mathematician. The year was 1971. I was 24 years of age.

CHAPTER 1

The Flood Plain

Tuscany, Italy, September 2002. Like many present-day travelers to Pisa, I took the train from Florence—a small commuter train of four carriages pulled by a noisy diesel locomotive, quite different from the sleek Intercity Express that had whisked me southward from Trento. Even late in the season, the train was crammed with tourists, many of them young people carrying backpacks. Everyone was talking loudly to make themselves heard over the noise from the engine. In my carriage I heard Americans, British, Australians, Germans, French, Scandinavians, and Japanese. A port in the Roman era and a major Mediterranean trading hub in medieval times, Pisa clearly is still an international destination, though these days the main cargo seems to be foreign tourists.

Once the train had left Florence behind, the journey became spectacular, winding its way through the beautiful rolling hills of the Chianti wine region. On both sides of the railroad tracks, the steeply rising slopes were covered with an irregular checkerboard of bright green vineyards, each one laid out with geometric precision. Occasionally, a field would stretch right down to the side of the tracks, giving the passengers a closer view. Now, in late summer, the vines were heavy with the ripening purple grapes that would soon be harvested to make the wines the region is so famous for.

Eventually, the hills gave way to a large flat plain, stretching all the way to Pisa and beyond to the sea. There had been heavy rains just prior to my visit to Italy, and as the train left the vineyards it began to rain once again. As the engine slowed down to arrive at our destination, I saw that the land on both sides of the tracks was still under a foot or more of water. The land here floods regularly, a lasting reminder of why Pisa had become a port in the first place: In Roman times, and earlier, this is where Pisa's harbor used to be.

By the time the train pulled up in Pisa, the rain had turned into a sustained, heavy downpour. The small, quaint, inexpensive hotel I had booked via the Internet was perfectly located for sightseeing, right in the center of the old medieval city, close to the river. Unfortunately, the railway station was not—it is a "Central Station" in name only. As I had experienced many times in New York City, when it rains in Pisa, everyone travels by taxi. As a result, the station taxi stand before me stood empty. I waited in line for an hour, with only my umbrella to keep me dry, before I was finally able to secure a ride. I soon began to wish I too had my belongings in a backpack, so I could have walked to my destination, as many of my fellow passengers did. It was a damp end to my journey, both literally and figuratively. Still, I was in Pisa at last, about to take the first step in what would turn out to be a seven-year quest to piece together the story of one of the most influential figures in human history, a medieval mathematician who, over the years, had become something of an obsession with me.

My visit had come about quite by chance. I had been invited to Italy to give an address at an international conference in Rome on the newly emerging field of mathematical cognition. I was asked to give lectures at several other universities as well—the industrial powerhouse of Torino in the northwest, the vacation destination Trento in the mountainous wine region in the northeast, the ancient university town of Bologna partway from Trento to Florence, and the spectacular Siena where, more than 20 years earlier, I had been a visiting professor for several weeks.

I had decided to take a two-day detour to Pisa in between my lecturing commitments in Bologna and Siena, in an effort to find out something about Leonardo Fibonacci, a mysterious thirteenth-century mathematician who apparently played a key role in the making of the modern world, and in whose mathematical footsteps I had, in one important respect, been treading for the past 20 years.

Was there enough information to write a book about him? No one else had written one, so I suspected there was not. On the other

FIGURE 1. This Leonardo woodcut provides one of only two images we have of Leonardo. There is no evidence either is more than an artist's conception.

hand, that yawning gap in the written history of science meant that Fibonacci was the most famous and accomplished scientist never to have been the subject of a biography. I wanted to give it a try.

My interest was certainly not that of the historian, for such I am not. I am a mathematician. What intrigued me about Leonardo was that significant similarity between our mathematical careers. I sensed a kindred spirit.

As I sheltered under my umbrella, waiting for a taxi, I reflected briefly on how different my mathematical career had been from the future I had envisaged back in 1968, when I completed my bachelor's degree at the University of London and headed off to the University of Bristol to begin work on my doctorate.

Back then, when I was starting out, the only thing I knew about Fibonacci was that he was the mathematician who discovered the famous Fibonacci sequence (he didn't—I was wrong), which I knew had deep connections to human aesthetics (it doesn't—I was wrong). It was much later that I discovered he was one of the most influential men of all time. And that his greatness lay not in his mathematical discoveries—though he was without doubt the strongest mathematician of his time—but rather in his expository power. He had the ability to take what were at the time novel and difficult mathematical ideas and make them accessible to a wide range of people. Moreover, he had the instinct to do it in a way that in present-day terminology would be described as a "good marketing strategy."

As a young graduate student, my role models were not the likes of Leonardo Fibonacci, but the mathematicians who had made major mathematical discoveries—more recent mathematical giants such as Leonard Euler, Karl Friedrich Gauss, Pierre De Fermat, and Kurt Gödel. Like many young people embarking on a mathematical career, I dreamed of joining the ranks of the greatest—of proving a major theorem or solving a difficult problem that had baffled the best minds for decades.

Some of my contemporaries succeeded. In 1963, only a few years ahead of me, the young American mathematician Paul Cohen solved Cantor's Continuum Problem, a puzzle that had resisted all attempts at resolution for more than 60 years. But as is true for the vast majority of mathematicians, eventually I had to settle for far less.

During the course of my career, like most of the world's 25,000 professional mathematicians listed in the *International Directory of Mathematicians*, I solved a number of minor problems and proved several respectable but largely unremarkable theorems. I taught at various universities, in Scotland, Norway, Germany, Canada, and the United States (where I moved permanently in 1987), and I wrote a number of textbooks for mathematicians and students. Again, these are all fairly typical career moves for many academic mathematicians, though perhaps I moved around more than many and ended up writing more books than most.

But along the way, almost by accident, I discovered another talent, perhaps my true calling: an ability to explain often obscure, advanced mathematical ideas to a general audience. I found that, through my words, I could make mathematics come alive for others not versed in the subject.

An unplanned sequence of events resulted in my discovering this ability and thereby embarking on a second career path as a public expositor of mathematics. In the early 1980s, having returned to the UK after four years in Norway and Germany, I grew increasingly frustrated by the fact that magazines and newspapers often carried articles on science—biology, physics, chemistry, and so on—but hardly ever on mathematics. On the few occasions when they did cover mathematics, they did so badly, often getting the main idea entirely wrong. In March 1983, I decided to do something about the situation, so I wrote a short piece and sent it in to the British national newspaper the *Guardian*.

It was an April Fools joke, to be published on April 1. I described some mathematics that, while true, was so counterintuitive, most readers would note the date and assume it was a spoof—and in so doing they would fall victim to the real joke: The article was true.

A few days later, the science editor, Anthony Tucker,[1] phoned and informed me that they could not publish it. "But," he said, "I like your style. You seem to have a real knack for explaining difficult ideas in a way ordinary people can understand."

Tucker encouraged me to try again, and my second attempt was published in the *Guardian* on May 12, 1983. Several more pieces also made it into print, eliciting some appreciative letters to the editor. As a result, when the *Guardian* launched a weekly, personal computing page later that year, it included my new, twice-monthly column *Micromaths*. The column ran without interruption until 1989, when my two-year visit to Stanford University in California turned into a permanent move to the United States.

I soon discovered that I liked my new role of "expositor." I have always been passionately interested in all aspects of mathematics, and never liked the fact that so many people have a completely false impression of this wonderful subject. Most people think that mathematics is just about numbers, but that's not true at all. Yes, numbers play an important role in the subject, but mathematics is not about counting. It's about pattern and structure. It's about the hidden beauty that lies just beneath the surface of the everyday world. I relished the challenge of constantly trying to find ways to explain new developments in advanced mathematics to

---

[1]   Tucker passed away in 1998.

the lay readers of my column. The frequent appreciative—and occasionally baffled—letters I received from readers further fueled my commitment.

Encouraged by the success of my column, I began writing books and articles for a general readership, including some for the business world. I also gave lectures to lay audiences and started to make occasional appearances on radio and television. From 1991 to 1997, after moving to the United States in 1987, I edited *FOCUS*, the monthly magazine of the Mathematical Association of America, and since January 1996 I have written a monthly column, "Devlin's Angle," for the MAA's Web magazine, *MAA Online*. (The column is now in blog format.)

Early in 1995, I got a break that led to my becoming a regular contributor to primetime national radio in the United States, with the media identity "the Math Guy." I got a telephone call one day from National Public Radio's Saturday morning news magazine show *Weekend Edition*. The host, Scott Simon, wanted to interview me about the solution to the 350-year-old problem known as Fermat's Last Theorem, which became a major news story after the Princeton mathematician Andrew Wiles had solved it a few months earlier.

Although Scott and I would not meet for many months—then as now, we record most of our interviews with me in a studio in California and Scott at the NPR studios in Washington, DC—we hit it off immediately over the air. Listeners loved our intimate, humorous banter—which from the start has been completely un-rehearsed and spontaneous. Many wrote in to the program to say so. Again, without any planning, I found I had another new role, this time a "radio personality," appearing on the show every few weeks. Eventually, I acquired my "stage name." The receptionist at the studio I used soon started to greet my arrival with "It's the math guy." I mentioned this to the *Weekend Edition* producer one day, and he replied, "Oh, that's what we put you down as on our scheduling board." And so the NPR Math Guy was born.

Each new step brought me further pleasure, as more and more people came up to me after a talk, or wrote or emailed me after reading an article I had written or hearing me on the radio. They would tell me they found my words inspiring, challenging, thought-provoking, or enjoyable. Parents, teachers, stay-at-home moms, business people, and retired people would thank me for awakening in them an interest and a new appreciation of a subject they had long ago abandoned for being either dull and boring or beyond their understanding. I came to realize that I was touching people's lives, opening their eyes to the marvelous world of mathematics.

None of this was planned. I had become a "mathematics expositor" by accident. Only after I realized I had been born with a talent that others appreciated—and that by all accounts is fairly rare—did I begin to work on developing and improving my "gift."

In taking mathematical ideas developed by others and explaining them in a way that the layperson can understand, I was following in the footsteps of others who had also made efforts to organize and communicate mathematical ideas to people outside the discipline. Among that very tiny subgroup of mathematics communicators, the two who I regarded as the greatest and most influential mathematical expositors of all time are Euclid and Leonardo Fibonacci. Each wrote a mammoth book that influenced the way mathematics developed, and with it society as a whole.[2]

Euclid's classic work *Elements* presented ancient Greek geometry and number theory in such a well-organized and understandable

---

[2] A close third, by my reckoning, would be Abū ʿAbdallāh Muḥammad ibn Mūsā al-Khwārizmī, whose ninth-century Arabic books on Hindu-Arabic arithmetic and on algebra were also written for a wide audience, though I did not know much about him back then. But because of the cultural stagnation that overcame the Arabic-speaking world in medieval times, and that continues to this day, it was left to Leonardo to make that body of knowledge available to the world.

way that even today some instructors use it as a textbook. It is not known if any of the results or proofs Euclid describes in the book are his, although it is reasonable to assume that some are, maybe even many. What makes *Elements* such a great and hugely influential work, however, is the way Euclid organized and presented the material. He did such a good job of it that his text has formed the basis of school geometry teaching ever since. Present-day high school geometry texts still follow *Elements* fairly closely, and translations of the original remain in print.

Because geometry was an obligatory part of the school mathematics curriculum until a few years ago, most people have been exposed to Euclid's teaching during their childhood, and many recognize his name and that of his great book. In contrast, Leonardo of Pisa and his book *Liber abbaci* are much less well known. Yet their impact on present-day life is far greater. *Liber abbaci* was the first comprehensive book on modern practical arithmetic in the Western world. While few of us ever *use* geometry, people all over the world make daily use of the methods of arithmetic that Leonardo described in *Liber abbaci*.

In contrast to the widespread availability of the original Euclid's *Elements*, the only version of Leonardo's *Liber abbaci* we can read today is a second edition he completed in 1228, not his original 1202 text. Moreover, there is just one translation from the original Latin, in English, published as recently as 2002.

For all its rarity, *Liber abbaci* is an impressive work. Although its great fame rests on its treatment of Hindu-Arabic arithmetic, it is a mathematically solid book that covers not just arithmetic, but the beginnings of algebra and some applied mathematics, all firmly based on the theoretical foundations of Euclid's mathematics.

I will describe my own reaction on first reading *Liber abbaci* in my fairly lengthy chapter 10 of this text, and, for readers who want to know more, I provide a summary of the entire contents of *Liber abbaci* in the appendix. For now, however, let me set the scene for the story I will tell by giving you the overall flavor of Leonardo's book.

Leonardo established a range of general methods for solving arithmetical problems (some using the geometric algebra of Book II of *Elements*), providing rigorous proofs to justify the methods, in the fashion of the ancient Greeks.

In particular, he explained—and provided justification for—some non-algebraic methods for solving problems that were well known in the medieval world, such as the checking procedure of "casting out nines," various "rules of proportion," and methods called "single false position" and "double false position," none of which are taught to today's calculator-carrying students. Indeed, these methods had fallen out of fashion by the time I learned arithmetic in the 1950s, a decade before the arrival of the digital desk calculator! (I did look up some of those methods when I was carrying out my Leonardo research, but I have already forgotten what they are.)

The real impact of the book came from its examples. Leonardo included a wealth of applications of mathematics to business and trade. These include conversions of money, weight, and content, methods of barter, business partnerships, and allocation of profit, alloying of money, investment of money, and simple and compound interest.

Presumably to add some variety and keep his readers' engagement, he also peppered his account with a number of highly artificial, cutely formulated, "fun" problems designed to illustrate various aspects of the mathematics he was describing.[3] For some of these "fun problems" he presented ingenious solutions that may have

---

[3]   Including "fun" problems is a literary device that I, and all other mathematicians who write for a broad audience, use frequently.

been of his own devising. One of his fun problems would prove to be forever identified with the name Fibonacci.

Incidentally, the unusual spelling of abbaci, with two b's, seems to have been introduced by Leonardo, to distinguish it from the name for the various kinds of devices merchants used to perform their calculations. For what *Liber abbaci* described was how to compute without using such aids. (It is definitely not the "book of the abacus" in the modern interpretation of the word "abacus"—with one b.)

After completing the first edition of *Liber abbaci*, Leonardo wrote several other mathematics books, his writing making him something of a celebrity throughout Italy—on one occasion he was summonsed to an audience with the Emperor Frederick II. Yet very little was written about his life.

In 2001, I decided to embark on a quest to try to collect what little was known about him and bring his story to a wider audience. My motivation? I saw in Leonardo someone who, like me, devoted a lot of time and effort trying to make the mathematics of the day accessible to the world at large. (Known today as "mathematical outreach," very few mathematicians engage in that activity.) He was the giant whose footsteps I had been following.

I was not at all sure I could succeed. Over the years, I had built up a good reputation as an expositor of mathematics, but writing a book on Leonardo would be a new endeavor. I would have to become something of an archival scholar, trying to make sense of thirteenth-century Latin manuscripts. I was definitely stepping outside my comfort zone.

The dearth of hard information about Leonardo in the historical record meant that a traditional biography was impossible—which is probably why no medieval historian had written one. To tell my story, I would have to rely heavily on the *mathematical* thread that connects today's world to that of Leonardo—an approach unique to mathematics, made possible by the timeless nature of the discipline. Even so, it would be a stretch.

In the end, I got lucky. Very lucky. And not just once, but several times. Three of my lucky breaks—and they were big ones—occurred very early on in the project.

My first stroke of luck, the biggest of all, came my way just as I was embarking on my quest. In 2001, an Italian historian of medieval mathematics at the University of Siena, Professor Rafaella Franci, was commencing the first-ever study of a late thirteenth-century manuscript in the collection in an archival library in Florence. Franci's analysis eventually determined (and other scholars subsequently confirmed) that the manuscript provided the long sought-after "missing link" to prove that Leonardo, and in particular *Liber abbaci*, was a major trigger for the arithmetical and financial revolution that began in Tuscany not long after the book's appearance, and in due course spread throughout northern Europe—all of which more anon.

As a result of that good fortune, when my historical account *The Man of Numbers: Fibonacci's Arithmetic Revolution* was published in 2011, I was able to compensate for the unavoidable paucity of information about Leonardo's life with the first-ever account of Franci's discovery showing that my medieval role-model expositor had indeed played a pivotal role in creating the modern world.

My second stroke of luck was the publication, in 2002, of a complete English translation of *Liber abbaci*, the first and hitherto only translation of the classic work into a modern language. This meant I was spared the task of brushing up on my school Latin in order to make sense of Leonardo's writing, and could focus instead on the mathematical content. Franci told me of this soon-to-be-published book when I first met her earlier in 2002, and I immediately pre-ordered a copy on Amazon.

My third lucky break occurred in 2004, when William Goetzmann, a professor of finance at Yale University, having also obtained and studied a copy of the new English translation of *Liber abbaci*, published a lengthy article titled *Fibonacci and the Financial Revolution*,[4] in which he analyzed Leonardo's text to show how essentially all present-day financial instruments have early counterparts there. (Combined with Franci's discovery, this provides compelling evidence that the modern financial world does indeed trace back to Leonardo.)

The Tuscan spring rain was pounding the railway station forecourt as hard as ever. Still, the queue of travelers in front of me waiting for taxis was getting shorter, as one by one a taxi would appear and whisk them away to their homes or hotel rooms. Soon it would be my turn. Then, tomorrow, I would really begin my quest to discover as much as I could about Leonardo.

Pisa. This is where it all began. This was where, exactly 800 years earlier, in 1202, Leonardo completed *Liber abbaci*—a book that, quite literally, would change the world. The task I had set for myself was to tell the world his story.

---

[4]   Published in Goetzmann and Rouwenhorst, 2005, pp. 123–43.

CHAPTER 2

The Manuscript

Pisa, Italy, 1202. The young man in his late twenties put down his quill pen and watched as the ink slowly dried on the page in front of him. The manuscript pile on one side of his desk stood some three inches high, and numbered over 400 pages.

The man had started the project some years earlier, when, as a teenager, he had left his childhood home to join his father, Guilichmus, or Guilielmo (William), Bonacci, a prosperous Pisan merchant who had recently been posted to the southern Mediterranean city of Bugia to serve as a trade representative and customs officer. There, in Muslim North Africa, the young man, Leonardo Pisano (Leonardo of Pisa), encountered Arab merchants and scholars who had revealed to him a wonderful invention, one that he was certain could transform the world.[1]

---

[1] Different people use the word "Arab" to mean different things. In this book, I use it in the commonly accepted scholastic sense to mean the peoples whose primary business or cultural language was Arabic—just as we speak of "the Greeks" to refer to the peoples whose primary cultural language was Greek. Used in this sense, the term Arab includes peoples from many nationalities, mostly, but not all, Muslim.

The men of learning in Bugia who showed Leonardo the invention were not its discoverers; they merely passed it on. The invention itself was much older, having its origins in India some time before 700 CE. Arab traders had transported it northward across land to the shores of the Mediterranean.

Leonardo took the page he had just written and placed it carefully on top of the pile. Now the book had a title. He read through that top page again:

Here begins the Book of Calculation
Composed by Leonardo Pisano, Family Bonaci,
In the Year 1202

(Like all European scholars at the time, Leonardo wrote in Latin. The title of his book in Latin was *Liber Abbaci*.)

The invention Leonardo had written about was a remarkable new way of writing numbers and calculating with them. So remarkable and complete would be the transformation of human life brought about by this new system that generations that came long after him would eventually take it for granted, relegating numbers and arithmetic to a tedious but necessary routine chore that had to be learned by repetition (or, for later generations still, left to machines). It is the system for writing numbers and doing basic arithmetic that is taught in schools today the world over. By making that system available to western European businessmen, Leonardo's *Liber abbaci* would play a major role in creating the modern world.

Part of the reason for the huge influence of *Liber abbaci* surely was circumstantial, as were the events that led Leonardo to write it in the first place.

To be born in Pisa in the twelfth century was to be born at the hub of the Western world. And to grow up in a Pisan merchant family was to be a member of what was then the most important

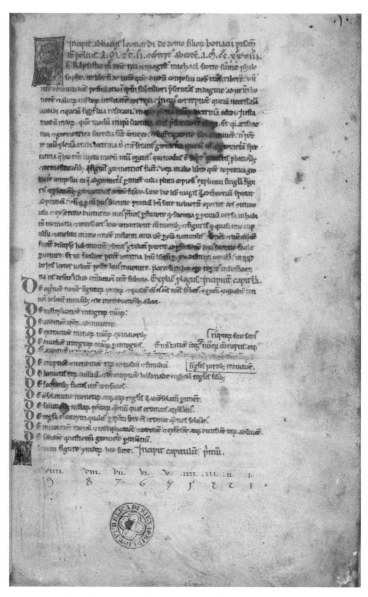

FIGURE 2. The first page of *Liber abbaci*, from the copy in the Siena Public Library.

sector of society. When Leonardo was born, Italy was a center of the vastly important and still rapidly growing international trade between the countries that fanned out from the Mediterranean Sea. Pisa, along with Italy's other maritime cities—Genoa to the north and Venice on the northeastern coast of Italy—dominated the trade, and their ships sailed constantly from one Mediterranean port to another. The merchants in those three cities were the key figures who were developing radical new ways of conducting international business, and thereby shaping the development of a new, more cosmopolitan world.

Yet time and place are only part of the picture. It usually takes true genius to see the greatness in the commonplace, and to recognize the enormous potential to change the world in what seems to most people to be a mundane or obscure idea. In the late sixteenth century, when Galileo Galilei gazed at the oil lamp hanging from the high ceiling in Pisa's Cathedral, swaying in the breeze, he made the key observation that led directly to the invention of the pendulum clock, the first accurate mechanical device to tell the time. In the seventeenth century, Newton saw an apple fall and realized there must be an invisible force—gravity—that not only governs all life on Earth but controls the motion of the entire universe. In the twentieth century, Albert Einstein wondered what it would be like to ride on the front of a beam of light, a seemingly childish question that led him to invent relativity—and with it nuclear power and the atomic bomb.

The pendulum, gravity, relativity—all of them single ideas, in retrospect simple ideas—that changed the world. So too with numbers. When the young Leonardo was living for several years in North Africa, he observed Arabic merchants using the system for writing and computing with numbers that they had acquired from the Hindus, and was convinced of the enormous potential of this invention, in particular for global commerce. As a result of that insight, he spent the next several years of his life writing an enormous instruction manual for using the new system.

Though Leonardo was born into a wealthy family with influential friends, and was well known for *Liber abbaci* and several other books when he died, there are almost no historical records relating to his life. While we know he was born sometime around 1170 CE, we do not know the exact year, and we are not completely sure where. Most likely it was Pisa, which is where he spent most of his childhood. According to the custom for naming at that time, in his later years, after he had become famous, he would have been known publicly as Leonardo Pisano. In 1838, the historian Guillaume Libri gave him the nickname "Fibonacci," a contraction of the Latin phrase *filius Bonacci* that Leonardo used to describe himself in the introduction to his book. (Although the literal meaning of *filius Bonacci* is "son of Bonacci," Bonacci was not the name of Leonardo's father, so the phrase should perhaps be translated as "of the family Bonacci.") The last contemporary reference to Leonardo was in 1240, in Pisa, but we have no idea how much longer he lived, or where or how he died.[2]

When I first became interested in Leonardo, I was surprised to find out that he had been largely forgotten just 200 years after he died. Yet, during his lifetime he had become famous throughout Italy and was honored by his native Pisa. But, as I learned from the medieval scholars I got to know, this was not an unusual occurrence, given the practice of the time. Other than the nobility, few people had anything recorded about them, even those who had achieved great things. The fame Leonardo enjoyed in his lifetime was because of his books, *Liber abbaci* among them. He was known as a brilliant mathematician and expositor of mathematics, and, later in life, a respected public servant. Clearly he was an accomplished

---

[2]   See chapter 12.

man. But no one at that time had any idea that their compatriot had made a unique contribution that ultimately would change the world. That was for history to judge.

Far more surprising—and I realized this only much later, after I had learned a lot more about the man—was that history did not make that judgment about Leonardo's greatness until the 1960s, and was able to make it with complete confidence only at the start of the twenty-first century, one year after I embarked on my quest to tell his story.

So it was that, with no realization in the thirteenth century of what Leonardo's work would lead to, after one or two generations, virtually all mention of Leonardo of Pisa in the historical record dried up, and his name did not appear in any book on the history of science or mathematics for 400 years.

Then, in the late eighteenth century, an Italian mathematician, Pietro Cossali, came across a single reference to Leonardo in an early printed text by the famous Italian mathematician Luca Pacioli, *Summa de arithmetica geometria proportioni et proportionalità,* published in 1494.

Fortunately for history, Pacioli listed his sources, with Leonardo Pisano prominent among them, declaring in the book's introduction:

> And since we follow for the most part Leonardo Pisano, I intend to clarify now that any enunciation mentioned without the name of the author is to be attributed to Leonardo.

That reference led Cossali to wonder why Pacioli was famous while the man whose work he had apparently built on so extensively was unknown. When he followed up by doing some research into this mysterious "Leonardo Pisano," he came to understand what a major role the thirteenth-century Pisan played in the spread of modern arithmetic. Leonardo's 400 years of historical anonymity were over.

In 1838, following the publication of Cossali's work in 1797–1799, the French historian Guillaume Libri gave Leonardo the manufactured surname Fibonacci. Then, in the 1870s, another Frenchman, the mathematician Edouard Lucas, assigned the name "Fibonacci sequence" to a fascinating number sequence that arose when you tried to solve one of the more recreational problems Leonardo had included in *Liber abbaci*. Soon thereafter, the name Fibonacci not only became widely known, but eventually achieved the status of a cultural icon, its spread being driven by interest in that particular number sequence.

Yet, were it not for that one reference to Leonardo by Pacioli, in all likelihood no one today would have been able to piece together the enormous role the Pisan played in human history. For what that reference did—coming from an accomplished, highly respected mathematician—was raise the possibility that Leonardo *might* have been the one who set the arithmetical revolution in motion.

And make no mistake about it, what followed the appearance of *Liber abbaci* was a revolution. (I should stress that I use the word "followed" here to mean "occurred soon afterwards." Establishing causation proved to be a difficult matter, one that Franci's discovery finally brought to a positive conclusion in 2003.)

The nature and scope of this revolution came to light little more than 50 years ago, with the discovery—in the many Italian archives of medieval manuscripts—of an extensive genre of short, handwritten textbooks in "practical arithmetic" (also called "commercial arithmetic"), each occupying a hundred leaves or so, with text on both sides.

The existence of this extensive genre of manuscripts was completely unsuspected until the Italian historian Gino Arrighi came across them and started to publish transcriptions of their contents in the 1960s.[3] Then, in 1980, the American historian Warren Van

---

[3]   Arrighi, 1964, 1967, 1973, 1987.

Egmond assembled and published a catalog of more than 250 of them, stretching from the late thirteenth century to the end of the sixteenth.[4]

Historians refer to these manuscripts as *libri d'abbaco* ("abbacus books"), or *trattati d'abbaco* ("abbacus tracts"). It is highly likely that they numbered into the thousands (see momentarily).

Almost all abbacus books were written in vernacular Italian, usually in the local dialect of the author. They varied considerably in length and mathematical quality. The earliest were handwritten manuscripts, but after the invention of the printing press in the fifteenth century they became one of the first recognizable literary genres, with some abbacus books becoming early best sellers. More than 400 such texts, stretching over 300 years, survive to this day. This rapid proliferation of texts provides dramatic testimony to the importance people attached to learning the new arithmetic.

A typical abbacus tract would begin by explaining how to write numbers using the ten digits 0 to 9, how place-value works, and how to calculate with whole numbers and fractions. From there, it would go on to provide a selection of worked examples, mostly practical business problems. Typically, the author also provided multiplication tables and tables of square roots to facilitate the solution to more difficult problems. Overall, the problems tended to be simple to state, and produced fairly short, straightforward solutions.

Most abbacus books did not bear their author's name. They were often illustrated, and some carried annotations saying they were presented as gifts to patrons and important merchants. It is evident that they were written for a local audience, since the monetary problems they presented were usually expressed in terms of the currency of the local town or region. The author often began by promising to explain "the art of the abbacus as it exists in the town of . . ."

---

[4] van Egmond, 1980.

Particularly in the case of the more poorly written books, we don't know why they were written, or for whom. It is likely that, in many cases, they were written as personal study notebooks, not intended for use by anyone but the anonymous writer. After all, in an era when books had to be copied by hand, the most efficient (and arguably the only effective) way to learn practical arithmetic was to locate a local copy of a book on the topic, make your own copy (by hand, of course), and then work through it. That many of the abbacus books were created and used in just that way is clear from their appearance. The pages of carefully written main text are adorned with scribbled calculations, marginal notes, and diagrams, sometimes in a different colored ink.

This, incidentally, is why we can only guess at how many such manuscripts were written. The odds are clearly stacked against a personally written, teach-yourself-arithmetic notebook surviving 800 years. Yet 400 of them did just that. If, say, as many as one in ten were to survive, that would mean 4,000 must have been written. And one in ten strikes me as a very optimistic hypothesis for the survival over 800 years of a handwritten, personal notebook.

The disparity between the main text and the annotations found in many abbacus books, together with some significant copying errors, suggests that in many cases the manuscript's author had very little mathematical skill or understanding.

Others, though, are more scholarly, and may well have been written by teachers to use in classes on practical arithmetic. For, in addition to the sudden appearance of the abbacus books, the decades following the appearance of *Liber abbaci* also saw the growth, throughout Italy, of "abbacus schools" (*scuole d'abbaco* or *botteghe d'abbaco*), where young children, and perhaps adults, were taught practical arithmetic. Of course, along with the schools came perforce a whole new profession, that of "abbacus teachers" (*maestri d'abbaco*).

There was, then, a huge demand for books and courses in practical arithmetic, starting in the Pisa area in the decades following the appearance of *Liber abbaci*, and eventually spreading throughout Italy and beyond, into Mediterranean and Northern Europe.

Given the timing, and the geographic origin, of these new instructional manuscripts, it is hard to resist the conclusion that the appearance of *Liber abbaci* was the cause—the "gunshot that started the revolution."

In terms of the mathematics covered, the methods in the abbacus books could, for the most part, all be found in *Liber abbaci*, though some texts included a chapter on geometry, which Leonardo covered in a separate book, *De Practica Geometrie*. But, while *Liber abbaci* and *De Practica Geometrie* were lengthy, scholarly texts, written in Latin, the abbacus books were, as I noted earlier, far shorter, written in vernacular Italian, with much simpler examples than those Leonardo presented in his two masterpiece volumes.

You might think that some of the more mathematically able, early abbacus book authors took *Liber abbaci* (and perhaps *De Practica Geometrie*) as their starting point, and wrote their own simplified versions. But there is a problem with this conclusion—or rather, until recently there was.

For one thing, there is considerable overlap among all the abbacus books—in arithmetical content, book structure and organization, mathematical style, and example problems—showing that they were all produced by one author copying from another, with at most superficial changes. On the other hand, in terms of exposition, none of them had virtually anything in common with *Liber abbaci* (or *De Practica Geometrie*).

Yet they do all appear to stem from a common root. Here is how we know that. Because the abbacus authors used local currencies, with current value in their worked examples (and in some cases local weights and measures as well), present-day scholars have been able to trace the extent of the abbacus books, both geographically and over time. That enabled them to compare one abbacus text with another, making it possible to construct an ancestral tree of books, showing how the genre grew by way of localized and contemporized copying. When you trace that tree backwards in time, you cannot escape the conclusion that the entire genre began with a single "Abbacus Eve," the mother of all abbacus books.

But, since none of the abbacus books had anything in common with *Liber abbaci* (or *De Practica Geometrie*), that would be true of Abbacus Eve as well. So who wrote Eve? Whoever it was did so before 1290 or thereabouts, since that is the date ascribed to the first abbacus book that has been found, and the evident lack of mathematical ability of its author makes it clear that *it* cannot have been Eve. (This is the book Franci examined in Florence in 2001–2003.)

To write Eve, the author had to be capable not only of reading Leonardo's two massive, scholarly, Latin tomes *Liber abbaci* and *De Practica Geometrie*, but of understanding their contents sufficiently well to be able to produce a greatly simplified account of the same material. Moreover, since the ancestral tree leads back toward Pisa, the author most likely lived somewhere in the vicinity of that city.

Based on the historical record, there is only one possible candidate: Leonardo himself. Another mathematician as accomplished as Leonardo would surely have left his own collection of writings.

Moreover, we know for certain that Leonardo *did* write a much simpler, shorter book on arithmetic. On a number of occasions, he referred to having written a *liber minoris guise* ("book in a smaller manner") to *Liber abbaci*. One such reference is in *Liber abbaci* itself;[5] another is in another of his books, *Liber quadratorum*, and there is a third reference in still another book he wrote, *Flos*.

In addition, the author of a later abbacus book refers to Leonardo's *libro di minor guisa o Libro di merchanti* ("book in a smaller

---

[5]   Page 154 of the Boncompagni edition of *Liber abbaci*.

manner or book for merchants"). That phrase "book for merchants" is significant, since it suggests that the *Libro di minor guisa* would most likely have comprised material from the first ten chapters of *Liber abbaci* together with parts of *De Practica Geometrie*.

The evidence pointing to Leonardo as the author of Eve was substantial. The one remaining problem was that no copy of Leonardo's simpler book had ever been found. So, whereas scholars were confident that it was, indeed, Leonardo who initiated the arithmetical revolution (and thereby the commercial revolution that followed—see momentarily), they did not know with much certainty what it looked like. Or rather, that was the state of affairs before Franci made her study of a particular abbacus book in the Florence archive.

Because of the date the abbacus manuscript was written— sometime around 1290—coupled with the results of a number of additional literary forensic studies carried out by other scholars since Franci published her conclusions,[6] we can now say with historical certainty that the manuscript Franci studied is indeed an early copy of Leonardo's simpler book on practical arithmetic, perhaps copied directly from Leonardo's original. (It was very clearly a "slavish copy," since the author's errors—presumably copying errors—and annotations indicate that it was done by someone who was not able to do much more than make a direct copy.)

It is possible that Leonardo wrote his simpler text in Latin and someone else translated it into vernacular Italian, but my money is on Leonardo having written it in Italian. His purpose in writing it was, after all, to make the "new arithmetic" as widely accessible as possible.

One obvious question that arises as a result of the discovery of the abbacus books is, what created the demand for all of these books and courses on practical arithmetic? That question turns

---

[6]   I summarize the more significant studies in *The Man of Numbers* (2011).

out to be easy to answer. The key factor was the rapidly grow-ing importance of arithmetic in the fast-changing world of the thirteenth century, particularly in the Italian regions of Tuscany, Lombardy, and to some extent Umbria.

The century after the appearance of *Liber abbaci* saw the birth of the modern world of global trade, commerce, and finance, with the introduction and development of, in particular, banking, insurance, and double-entry bookkeeping, together with the growth of ever larger trading conglomerates. All of those activities depend on the ability of an efficient way to perform arithmetical calculations that can be mastered by all.

Prior to the adoption of the Hindu-Arabic system of arithmetic, traders used one of two methods to carry out their calculations: finger arithmetic or a mechanical abacus. The former was a highly sophisticated system using all the fingers and thumbs of both hands—the general English term is "digit," of course, from the Latin *digitus* meaning finger or toe. (This is how finger arithmetic led to our present-day use of the word "digit" for the basic number symbols 0, 1, 2, through 9.)

Ancient finger arithmetic was capable of representing numbers up to 10,000. As you can imagine, it required considerable training to master the system and a lot of practice to become fluent. Me-dieval textbooks, including some surviving early copies of *Liber abbaci*, often had an initial page with carefully drawn diagrams showing the positions of the fingers to represent various numbers and perform different operations.

In the Muslim world and much of Europe, the abacus was a flat board with ruled lines on which small pebbles were placed and moved around. This form of early calculation gave rise to our pres-ent word "calculus" for a system of calculation procedures (and hence led to the word "calculate" itself), since the Latin word for a pebble is *calculus*. The more familiar form of abacus now found in toy stores, with beads on wires instead of pebbles on a rules board, was used in China, where it was called a *Xuanpan*. As with finger arithmetic, calculating with an abacus required considerable practice to achieve accuracy and speed.

Traders became highly proficient with whatever method of calculation they used, so the two systems were efficient—at least for one-on-one trading. The significance of that last caveat is that neither method leaves a record of the calculation. One obvious consequence is that if either party suspected an error had been made, the only option was to repeat the calculation, and continue to do so until both traders were satisfied. But that was not the big issue. Another, less obvious drawback was far more significant. When the businessmen in twelfth-century Italy started to form trading empires, they had a clear need for an audit trail. An individual sitting in Pisa controlling a network of traders needed to be able to review the financial books on a regular basis.

Hindu-Arabic arithmetic, carried out on paper, was not only much easier to learn and master than finger arithmetic or use of an abacus board, it also left a clear audit trail that anyone could inspect and check for arithmetical accuracy. Almost certainly, then, two factors contributed to the rapid adoption of Hindu-Arabic arithmetic in thirteenth-century Italy: its relative ease of learning and use, and the automatic provision of a complete record of any calculation.

The latter factor surely explains why no commercial revolution followed the publication of a previous book on Hindu-Arabic arithmetic in the ninth century, by the Persian mathematician Abū ʿAbdallāh Muḥammad ibn Mūsā al-Khwārizmī, which Leonardo almost certainly consulted in writing *Liber abbaci*. (It is from al-Khwārizmī's name that we get today's word "algorithm" for a procedure to carry out a computation.)

In fact, al-Khwārizmī went on to write a second book describing how to do arithmetic in a more general fashion that can be scaled, with a word in that book's title giving rise to today's term for that more general method: "algebra," from the Arabic *al-Jabr*.

Though today, non-mathematicians tend to think of algebra as "arithmetic with letters," that's actually a significant misunderstanding. For one thing, algebra can be done without using symbolic equations. In the first ever algebra textbook, al-Khwārizmī wrote everything in prose. The approach using symbolic expressions was introduced much later, by the French mathematician François Viète in the sixteenth century, and should properly be called "symbolic algebra." As originally developed, the methods of "algebra" approach numerical problems in a different way than do the methods of arithmetic. In arithmetic, you *compute* with numbers to arrive at an answer to a problem. In algebra, you postulate the existence of the answer, giving it a name (commonly the letter x in algebra as presented today), and then reason logically to deduce what x must be. "Numerical forensics" would be a more helpful name for this approach. Because the reasoning is logical, involving general relationships between classes' numbers, rather than calculating with specific numbers, solutions tend to be very general, enabling them to be applied to whole families of numbers. Today's electronic spreadsheet is a digital implementation of algebra much as the electronic calculator is a digital implementation of arithmetic.

Thus, as a result of al-Khwārizmī's books, ninth-century Persian and Arab merchants had access not only to Hindu-Arabic arithmetic but also to the more powerful, scalable method of algebra their mathematical countrymen developed. The reason the commercial revolution did not emerge from Baghdad at that time, as it would four centuries later when those methods found their way to Italy, was that the commercial world had not yet developed sufficiently for the new methods to have a widespread impact.

New innovations typically start out as a way to improve the efficiency of a process that is already being done. Very often, however, the innovation leads to a new activity that had not been done earlier, which then often replaces the old activity. In the case of modern arithmetic and algebra, that second step was making use of those methods to conduct international commerce *at scale*. It fell upon the Italian business communities in thirteenth-century Tuscany and Lombardia to take that second step.

As we now know, Leonardo's accounts of modern arithmetic and algebra provided the spark that lit the fire of the modern commercial world. But his words did so only because the spark fell on a highly combustible business landscape! Putting efficient methods for arithmetic and algebra into the hands of anyone willing to devote a few weeks to master them opened the door for ambitious individuals to seek their fortune through trade and commerce. It was the first personal computing revolution, with no shortage of people every bit as eager to step forward, just as happened in the 1980s with the second personal computing revolution that emerged from California's Silicon Valley, when the purchase of an electronic computing device provided a similar promise of a new future.

In Leonardo's time, however, the perspective of seeing that massive commercial transformation was not available, and the role he played in it could be neither recognized nor appreciated.

Nevertheless, many of the abbacus books in the early generations carried an introductory proclamation that their exposition was "in the manner of Leonardo Pisano." Why was that?

For the first few abbacus books, the citation was surely because Leonardo's writings had been so well known. But over time, that proclamation became little more than a ritualistic convention, carried from one text to another, and it is likely that most abbacus authors had no idea who that mysterious individual was or what he had done. And, as the decades turned into centuries, the habit gradually died away altogether—except, however, for Pacioli, who made such a proclamation at the start of his 1494 text.

Pacioli's book was significant in ensuring that Leonardo would one day be credited as a major figure in initiating the modern world. Because Pacioli was a famous and respected mathematician of great ability, his endorsement of Leonardo was almost certainly not ritualistic, but carried weight. Moreover, his book was not handwritten but printed—which implied accurate reproduction, much wider distribution, and greater permanence.

Ironically, were it not for Pacioli's one brief introductory paragraph mentioning Leonardo in a printed book, the arrival of printing would likely have been the nail in the coffin of Leonardo's future place in history. For if we want to understand why Leonardo's name disappeared in the first place, we need look no further than the invention of the printing press around 1450. Manuscript texts on practical mathematics were among the first works to be put into print (in no small part due to the demand for practical arithmetic that Leonardo's work had created, as we have seen). Not surprisingly, printers put their effort into typesetting the most accessible and most up-to-date works, among which were many shorter, simplified manuscripts descended from Leonardo's works, each updated with the most recent prices and currency exchange rates. Leonardo's much longer, quarter-century-old originals had long been left to gather dust in a few archives. And there they may have remained, undiscovered, were it not for Pacioli citing Leonardo as the original source for material in his printed book—a reference that lay dormant for 300 years until Cossali came across it in 1797.

CHAPTER 3

First Steps

Were it not for the invitation to present some lectures in Italy, I am not sure I would have actually embarked on my quest to discover Leonardo's story. At least, I wouldn't have done so at that time, when my academic career was doing well and I had many interesting projects in the works. (Though I did not realize it then, my invitation at just that point was yet another lucky break that moved my Leonardo research forward.)

So little was known about the man, and the few things I did learn from the literature raised even more questions. (Most of the historical development I recounted in the previous chapter was unknown to me when I began my journey back in time.) Leonardo's life was a tantalizing mystery, with many hanging threads. My Fibonacci notebook was pathetically thin, with only a few of my initial questions answered:

> *Who gave him the nickname Fibonacci, and when?* Okay, I had found the answer to that one.
> *How did Leonardo come to write* Liber abbaci, *and what else did he do?* That one too I had answered.

Why was the number system that Liber abbaci *introduced to the west initially banned from commercial use in Italy? How long did the ban last? How and why was it lifted?* The historians had answered that one too.

Who carved the statue of Leonardo that I had seen in photographs that appeared in books and on websites, and where was it now? In principle, I could have found the answer to that one, but, being unschooled in the methods of archival research, until I started making research trips to Italy and speaking with the historians, I did not know where to look.

What did Leonardo look like? There is no evidence that the statue is anything other than a work of fiction, and the same is true for the one existing drawing of the man, which is also of relatively recent origin. Likely not a question that a professional historian would worry about, but having seen myself as following in his expository footsteps, it was one I frequently found myself musing over.

When did Pisa name a stretch of the riverside road that runs alongside River Arno (by the Giardino Scotto) Lungarno Fibonacci (Fibonacci Way)? Why did Florence name one of its streets the Via Fibonacci?

Why is Leonardo better known for a number sequence, named after him as the Fibonacci sequence, that can be found in the petals of flowers and the leaves of plants, than for the manuscript that changed history? (Many of the thousands upon thousands of hits you get when you do a Web search on the name "Fibonacci" relate to the Fibonacci sequence.)

Why are some present-day stock trading companies named after Fibonacci?

Why is Leonardo not as well known as Galileo, the man who gave the world modern science? (After a short distance, the riverside road known as the Lungarno Fibonacci becomes the much longer Lungarno Galileo Galilei.)

> *Why did a group of California mathematicians come together in*
> *1963 to form the Fibonacci Society, and why to this day does the*
> *society publish a respectable mathematics journal, the* Fibo-
> nacci Quarterly?
>
> *Did Leonardo marry? Did he have children? Is there a lost record*
> *somewhere that provides the dates or places of his birth and*
> *death?*
>
> *Above all, what kind of a person was he? What was his daily life*
> *like?*

Those were just some of the questions I formulated when I embarked on my quest.

I intended to carry out much of my research in a largely traditional fashion, talking with experts on medieval mathematics and
looking through various archives. But that would take me only so
far. It was clear that a significant piece of the story would have to
be gleaned from Leonardo's mathematical writings.

I suspect that many people who are not well versed in mathematics
may find it hard to understand how it is possible to gain insights
into a mathematician's mind (in my case, Leonardo's mind)
through his mathematical writing, as I was proposing to do. But
for a mathematician—particularly one such as myself, with similar
interests—this is not much more difficult than getting a sense
of composers through their musical works, of novelists through
their books, or of moviemakers through their films. In the case
of Leonardo, we are after all not talking about a few scribbled
calculations, but a massive body of creative work, written over
many years.

It is true that the result of my approach would be a recreation
of Leonardo through modern eyes. But actually, the same is true of
any historical account. Besides, what better way is there for us to
understand a man whose greatness and influence on the world lay
in his mathematical writings, than through those very writings?

The first real step, I eventually realized, would be to go to Italy and—taking advantage of the fact that much of the country is unchanged since medieval times—try to get a visceral, human sense of what life must have been like in Leonardo's time, so that I could start to flesh out the words he left us on his parchment manuscripts.

When that (all-expenses-paid) trip to Italy fell into my lap, my quest was really under way. This would be the first of many research visits I would make over the coming years. It would take me a decade to find answers to (almost) all of my questions.

The result of my research would finally be published as a book in 2011, with the title *The Man of Numbers: Fibonacci's Arithmetic Revolution.* But such was the magnitude of the task ahead of me that, when I set out on my quest in September 2002, most of what would constitute the book's main storyline was unknown to me. Indeed, the key historical manuscript that definitively tied everything together had only just been discovered, and not yet made public!

From the outset, I decided to keep a project notebook, so I would have a record not just of what I discovered, but how I discovered it. The notebook would chronicle my journey into the past, and perhaps serve as a source of factoids and anecdotes to use when I went on the inevitable author tour to promote my book. (My secondary career of a public expositor of mathematics had schooled me in the world of commercial book publishing, and I already had a series of popular books to my credit.) But, as I was selecting the material for what would be *The Man of Numbers*, I realized that, while my notebook began as a record of my research into Leonardo and *Liber abbaci*, it actually told a very different—and complete—story of its own.

For it turned out that I was not alone in my fascination with Fibonacci the man. The secondary story my notebook told was of five present-day individuals, myself included, who each sought to contribute in their own way, and for their own reasons, to getting Leonardo's story into the public realm.

There was Laurence Sigler, an American mathematician at Bucknell University in Pennsylvania, who labored for many years in his free time to translate *Liber abbaci* from Latin into English (the only translation of the text into a modern language), only to die shortly before completing the task.

There was Sigler's Hungarian-born widow, who taught herself the Latin, the mathematics, and the sophisticated mathematical typesetting skills necessary to complete her husband's work after he had passed away.

There was William Goetzmann, a distinguished professor of finance at Yale University, who, using Sigler's translation, was able to trace most of modern finance back to Leonardo's book.

And there was Rafaella Franci, who I had arranged to meet at the end of my short visit to Italy. Franci was a historian of mathematics at the University of Siena, and it was she who, shortly before my visit, had made the crucial archival discovery that brought together all the threads in the story of *Liber abbaci*. (Though it would be two years later before she told me of that discovery, after her paper had gone through academic peer review and been published in a journal.)

So rich were their stories that, what began as a record of my research activities, grew naturally into an account of how five individuals unraveled, and then told, different parts of the story of one of the most influential scholars the world has ever known.

The Man of Numbers tells the thirteenth-century story of Leonardo, and in particular his seminal book *Liber abbaci*. The pages you have in front of you now tell the story of the five committed individuals whose collective efforts ensured that Leonardo occupies the place in the history of human ideas that his work rightly deserves.

Turning my record into this new book has been a much greater stretch than *The Man of Numbers*. Like most non-fiction authors, I comfortably adopt the mantle of an outside, third party, writing about mathematics, science, engineering, and the people who have made the great advances. This new book would have to be in the first person. I would be describing things I did, feelings I had, conclusions I drew. I knew I would be opening my inner self up more than in any previous book. It would unavoidably cross several different literary genres. As my agent of many years (Ted Weinstein) can affirm, finding the right voice to tell this story has taken a lot of trial-and-error, and more fundamental re-writes than anything else I have ever written—the last rewrite in response to suggestions by my editor at Princeton University Press, Vickie Kearn.

Over the many years of my Fibonacci project, I have come to love the mythical image of Leonardo I created in my mind through reading his many words, by walking where he would have walked and seeing buildings he would have seen. Would I do him justice? Would I do justice to to the American who spent the last years of his life struggling to complete an English translation of *Liber abbaci*, or to his widow who completed the work after his cancer won the battle? The story told in this book is not something I learned about and studied, nor is it a tale I made up. It is an account of events I experienced over more than a decade of my life. There is a lot of me on the pages of this book—and that is a new experience for me.

If you ever wondered how a history book comes to be written, my chronicle will give you one example. It is by no means a matter of spending hours in libraries, or sitting in front of computer screens, or conducting interviews, though all of those were involved. For me, at least, writing a history book has been a human journey, full of highs, lows, false starts, frustrations, unexpected turns, tragedies, amusements—and sometimes plain, old-fashioned luck.

But back on that rainy evening at the Pisa Central Station in 2002, all of those steps were ahead of me. I knew nothing then about those three other scholars who, like me, had developed an interest in Leonardo. The intertwining of our paths was yet to come.

At last, it was my turn to escape that shiny-wet station forecourt. I was at the head of the queue and the next taxi would be mine. It was still raining. I hoped the weather would be clear by the next day. For tomorrow my search for Leonardo would begin in earnest. I was about to take my first tentative steps as a mathematical historian.

CHAPTER 4

The Statue

PISA DAY 1. During the night, the rain had stopped and the weather had indeed cleared up. With my time in Pisa so limited, I had worked out a schedule in advance. Like any tourist, after indulging in a post-breakfast cappuccino—this was Italy after all—I would spend my first morning on the obligatory tour of the Piazza dei Miracoli (the Square of Miracles), where the famous Leaning Tower is located. In the afternoon, I would visit the city's two Leonardo memorials that I had read about in my prior research, a street sign and a statue. (There is a third memorial I learned about later, that I would visit on a subsequent trip.) Then, the following day, I planned to take a walk around the rest of the city, to try to recreate in my mind the picture of daily life in Leonardo's time that I had gleaned from articles and books.[1] (The heavy research would come later.) In many parts of the world, this would not be possible, but in Italy, large parts of many of the cities date back to

---

[1]　In particular, the pamphlet by Gies and Gies (1969), long since out of print, but I was able to track down a copy.

medieval times, with buildings much as they were in the twelfth and thirteenth centuries.

For example, visitors who stroll around today's Pisa will come across tall rectangular towers, built of stone or brick, rising three or more stories high, that date back to Leonardo's time. With constant feuding between rival families, the tower provided any Pisan family of means with a refuge as much as a home. The ground floor was often a shop or a storeroom for oil, wine, tools, and supplies; the second floor served as the main living area, and perhaps a bedroom. The kitchen was usually on the top floor, to allow for smoke to escape easily. The common Pisan boast that the city had 10,000 such towers was surely a huge exaggeration, but as a child in a wealthy merchant family, Leonardo almost certainly grew up in such a building.

I set off early for the Piazza dei Miracoli. So too did hundreds of other visitors to the city. Most of them had come to Pisa to see, and perhaps to climb, the Leaning Tower. This 179-foot tall bell tower, begun in 1173, has become one of the most famous monuments in the world, and is the unofficial but universally recognized emblem of Italy.

The Piazza dei Miracoli is a complex of buildings in the northwest corner of the city that includes the cathedral, the baptistry, and the (leaning) bell tower. Construction had been underway for more than a century when Leonardo was born, and the cathedral and the baptistry were complete, although the baptistry dome would not be added for another hundred years. Work on the bell tower was just beginning when Leonardo was growing up. Walking around the square as a young boy, he would have seen marble blocks for the tower being hauled onto the site on heavy carts, having been transported to the city by barge from quarries in the mountains. He would have seen stonecutters give the blocks a final dressing before hoisting them into position with cranes and mortaring them together.

Just when the tower reached three of its intended eight stories, it started to topple to one side. In a valiant effort to avoid a complete collapse, Bonnano, the engineer in charge, started to make the upper stories slightly taller on one side to compensate for the lean. But the additional weight of the extra masonry on that side simply caused the foundations to sink even further. In the end, all work was suspended until an acceptable solution could be found. When the tower was eventually completed in the fourteenth century, it still leaned—as it does today.

At the time of my visit, a major engineering project had just been completed, reducing the tower's lean. But it still looks as though it ought to fall down—as one day it probably will.

Besides the Leaning Tower, most tourists also visit the two other famous buildings in the piazza—the cathedral, and the baptistry. But few of the many thousands who stroll the paths that cross the rich green lawns of the square, marveling at the white marble architecture, are aware that just a few paces away stands another monument to Pisa's glorious past. There stands the statue of the man who quite literally changed the course of the world, giving Europe—and much later, North America—a crucial key to all of modern science, technology, medicine, engineering, and commerce. Still, I can hardly blame the tourists for not knowing of him. Even among my colleagues in mathematics, few seem to regard Leonardo with quite the same fascination I had developed for the man. And none, as far as I know, had ever tried to find out everything about him, as I was now setting out to do.

Because of the crowds, the tour of the piazza took longer than I had anticipated, so it was already afternoon before I set forth in search of the Leonardo memorials. I had little idea of how difficult my search would be, nor of the degree to which today's Pisans are unaware that one of the most influential figures in human history grew up in the city.

My first Leonardo point of interest was a street in Pisa called the Lungarno Fibonacci. It was so named in a session of the Consiglio Comunale (Town Council) of Pisa on September 28, 1950, and is now one of nine streets in Pisa named after a historical figure sufficiently well known to be called "Pisano" (meaning "of Pisa"). *Lungarno* is a reduced form of the Italian term for "along the Arno," and the street runs alongside the south bank of the river at the eastern end of the city, adjacent to a delightful, if slightly run down, little park called the Giardino Scotto. According to some sources I had consulted, the park is named after Leonardo's friend Michael Scott, to whom he dedicated *Liber abbaci*, but I discovered that this is not the case. More reliable authorities say that the name comes from Domenico Scotto, a Livornese ship-owner who bought the property in 1798.

I took photographs of two Fibonacci street signs—one a rather ordinary metal plate on top of a pole by the bridge over the Arno, the other a splendid, carved stone slab, mounted high up on a brick wall covered with ivy—so high I had to reach up on tiptoes and hold the camera above my head to photograph it. Some passersby

FIGURE 3. A Leonardo Street sign in Pisa, photographed in 2003.

looked at me curiously, no doubt wondering why anyone would go to so much trouble to photograph a street sign.

The second memorial I wanted to see was a statue of Fibonacci. This proved harder to track down. I had read about it in several places, and had seen photographs, but sources differed as to its location. One website even claimed there are two statues, which I eventually learned is not true. After several hours of walking in the hot sun, from location to location in what seemed like circles, I still had not found it.

Most sources claimed it was located in the Giardino Scotto. I tried there first, when I visited the Lungarno Fibonacci, but could not find it. I asked the Giardino Scotto groundskeeper, who looked like he'd tended the park his whole life. Unfortunately, my Italian was poor and he gave me a puzzled look, seemingly surprised that a tourist was talking to him. I would eventually learn that the statue used to be there, but was moved in 1990.

One Web source said that the statue was located in a cemetery adjacent to the Piazza dei Miracoli, so I traipsed back across town to where I had spent the morning. I discovered that there is a cemetery next to the Piazza dei Miracoli, a small Jewish cemetery at the northwest corner of the square. But that did not seem to be a very likely location in which to find a memorial to the presumably Catholic Leonardo. And indeed, it is not there. So where was this statue?

I spent the next two hours walking round the outside of the piazza, going along every street, looking into gateways and peering over walls, searching for another cemetery, perhaps even a small, private one, but to no avail. Clearly I needed help.

What better place to determine the location of a famous monument than in an official Tourist Information Office?

"Where is the statue of Leonardo of Pisa?" I asked. The rather stern-looking lady sitting behind the desk gave me an irritated

"Don't bother me" look. I felt a twinge of guilt for disturbing her. Still, this was an official City of Pisa Tourist Information Office, located on the edge of the Piazza dei Miracoli, directly across from the famous Leaning Tower, and for today I was a tourist. Eventually she got around to replying.

"You mean Leonardo Da Vinci." It was not a question. Her tone was that of a schoolteacher addressing a pupil who has not done his homework.

"No," I replied patiently, "Leonardo of Pisa."

She gave me one of those withering looks that Italian and French people reserve for tourists who come on vacation largely unaware of the fine cultures of those two countries. Only I was not totally ignorant. I'd been doing some preliminary research about Pisa, and in particular on one of its greatest citizens ever.

"Leonardo Da Vinci," she repeated in impeccable English, this time speaking very slowly, as if to a small child, and putting great emphasis on the word "Vinci."

She clearly regarded me as an ignoramus who had not even heard of one of Italy's (though not Pisa's, it should be said) most famous artists.

"No, Leonardo of Pisa," I said again, stressing "Pisa" this time. "Fibonacci," I added, "mathematician."

Her exasperation showed. "There is no Leonardo of Pisa," she declared firmly. "No statue."

With that, she looked down at her desk, seemingly more interested in filling in some forms than answering my question. It was obvious that she was not going to look up again until I had left, which I did.

Dusk was approaching as I stepped back into the Piazza dei Miracoli, and I was starting to get frustrated. I'd spent almost the whole day in Pisa and all I had to show for it were some photographs of two street signs. I took a second, more thorough look at one of the tourist information signs posted around the square. There are, I discovered, not three but four buildings that make up the religious complex of the Piazza. In addition to the cathedral, the baptistry, and the bell tower, there is a fourth building, the Camposanto, which was started later, in 1278, after the other three buildings were essentially completed. Its English name, the tourist sign said, was Monumental Cemetery. "Cemetery!" I suddenly had one of those "Aha!" moments. This has got to be it, I thought.

Compared to its three sisters, the face the Camposanto presents to the outside world is unremarkable. Apart from an ornately carved Gothic tabernacle that rises up above one of the two large metal doorways that open out toward the cathedral, all the visitor sees from the piazza is a long, low, clean, white stone wall. The building was extensively renovated and cleaned after severe damage during the Second World War. Its exterior looks very modern, quite unlike the other monuments in the piazza. I had not given it a second look on my earlier tour. Now I did.

With only a few minutes before the museum closed for the day, I bought an entrance ticket and raced inside, not sure what I would find. I discovered that the Camposanto has an elegant, understated beauty that it keeps hidden from the outside world. It faces inward, with four cloistered walkways looking onto a long rectangular lawn. I turned left and walked around the western end. And there, facing me, at the far end of a long, marbled corridor in front of the eastern wall, was the imposing statue of Leonardo Fibonacci. It glistened pure white, standing out from the shadowy corner behind it, quite unlike the grimy, dark, and somewhat foreboding, gothic-looking statue in the Giardino Scotto I had seen depicted in published photographs.

A moment later, there I was at last, staring up at Leonardo of Pisa. Or was I? Did the sculptor Giovanni Paganucci have access to a reliable source showing Leonardo's appearance, or was the statue purely a product of his imagination? There is no known contemporary drawing of Leonardo, so the statue is most likely a work of pure fiction, but we may never know for certain. The one other image that purports to be of Leonardo is an engraving of the head and shoulders of a young man, often reproduced in textbooks, but it too is fairly recent. There is little resemblance between Paganucci's statue and the engraving.

Accurate or not, the white marble figure in the Camposanto portrays a tall, slim, handsome man, seemingly in his early forties. His face is sculpted with classic features, an angular face with a strong, narrow nose, a well-defined jaw, and a cleft in his chin. He stands erect, draped in a long, cascading tunic, with his head covered by a hood, giving him a vaguely clerical look. Two locks of carefully curled hair peek out from under his hood. He looks down from his plinth with a kind, scholarly expression, holding a book in one hand—*Liber abbaci*, perhaps? (Actually no. Though people did bind together manuscripts into a "book form" in Leonardo's time, they did not resemble modern books, like the one Leonardo's statue holds in his hand.) His other hand is stretched forward as if gesturing to emphasize a point to a student. I noticed that both hands had fingers missing, the one defect in an otherwise beautifully carved statue. I liked him instantly.

I looked down and read the inscription on the pedestal on which Leonardo stands:

A Leonardo Fibonacci Insigne Matematico Pisano del Secolo XII

The Italian was within my capacity. I translated it as: "To Leonardo Fibonacci, noted mathematician of Pisa of the Twelfth Century." I noticed that the date was a bit misleading. Although Leonardo was born around 1170, all of his mathematical publications were in the thirteenth century, starting with the first edition of *Liber abbaci* in 1202.

FIGURE 4. The author at the statue of Leonardo in the Camposanto in Pisa, 2002.

Far down at the rear end of the right side of the pedestal I saw a small inscription that reads:

G. Paganucci
Firenze 1863

With the light fading rapidly, I quickly pulled out my camera to take some pictures. At that moment, the fates conspired against me once more, as a group of elderly ladies turned the corner and gathered in front of me, the object of their attention being one of the gravestones that make up most of the pathway. They started to translate its Latin transcription aloud—slowly and intently. They gave no more than a passing glance at the statue towering above them, a memorial to a man who had surely had a far greater effect on their lives than the individual whose remains lay beneath their feet—whoever he was. After what seemed like an eternity, the group moved on to try their hand at another inscription, and I managed to take my photographs with just moments to spare before the last light finally went.

My search over, I left the building, surprising the attendant at the front gate, who, assuming that everyone had left long before, was preparing to lock up for the night. I was pleased that Leonardo is displayed so prominently. The statue's splendid location shows that some Pisans, at least, have recognized the importance of their thirteenth-century forebear who gave the west numbers—even if the lady in the City of Pisa Information Bureau, less than 50 yards away, knew nothing about the man or his statue.

Leonardo was now more than a mere name in a book. He had acquired a physical form in my mind. I was getting to know him.

But now I was faced with another question: What was the story behind the statue?

The October after I returned home from my trip to Italy, I wrote about my search for Leonardo's statue in my regular "Devlin's Angle" column.[2] After recounting my exchange with the woman in the Information Office, I ended my article with this request:

> In my brief time in Pisa, I was unable to find out anything
> definitive about the history of the statue of Leonardo. Who
> carved it and when? When exactly was it moved from the
> Camposanto to the Giardino Scotto and why? When was it

[2] http://www.maa.org/external_archive/devlin/devlin_10_02.html

renovated and moved back? If any reader knows the answer
to any of these questions, or has any other information about
the statue, I'd be curious to know.

A year and a half later, on March 20, 2004, I received an email
from a Sgr. Gian Marco Rinaldi in Italy, who said he had read, and
liked, my article. As a teacher of mathematics living in Torre del
Lago (home to the composer Giacomo Puccini, he proudly pointed
out), just 15 km outside Pisa, he had been inspired by my article
to do some local research into the Leonardo statue, to see what
he could find out. Here is what he told me.[3]

The initiative for creating the statue occurred not in Pisa but
in Florence, and was made by two politicians from ancient aris-
tocratic families in Tuscany: Baron Bettino Ricasoli and Marquis
Cosimo Ridolfi. In 1859, the Grand Duke was exiled and the following year Tuscany was annexed to the kingdom of Savoy, which
soon became the new unified Italian state. During the transition,
Tuscany was ruled by a provisional government. Ricasoli became
Prime Minister and Ridolfi served as Secretary for Education. They
both were active in promoting culture, including the founding
of a modern institute for advanced studies that later became the
University of Florence. On September 23, 1859, Ricasoli signed
a decree resolving that the State of Tuscany should finance the
carving of three statues, one each for the towns of Pisa, Lucca,
and Siena. Each was to commemorate an important local person;
the statue for Pisa was to be of Fibonacci. The decree cited him as
"the initiator of algebraic studies in Europe."

---

[3]   His principal source was an article by Rodolfo Bernardini, titled "Leonardo Fibonacci nella iconografia e nei marmi" [Leonardo Fibonacci in iconography and
in marbles], published in the magazine *Pisa Economica*, 1977 (n.1), pp. 36–39. The
article describes the memorial stone and the statue. *Pisa Economica* is published by
the Pisa Camera di Commercio, Industria, Artigianato (Chamber of Commerce).

The work was commissioned to Giovanni Paganucci, a sculptor in Florence, who completed the Leonardo statue in 1863. It was placed in the Camposanto, alongside the statues of many other prominent Pisans, and officially unveiled on June 17.

Leonardo remained in the Camposanto until 1926, when the then fascist authorities in Pisa removed that and two other statues from the Camposanto and placed them in three squares of the town, where they would serve as a testimonial to some of the great citizens that the city had produced. All three relocated statues were named "Pisano"—Leonardo plus the famous father and son sculptors and architects, Nicola and Giovanni Pisano. A sentence was added to the inscription on the pedestal of each statue that read:

Dall'oblio alla gloria per volontà fascista

Taking the translation of the Italian word per as "for," this translates literally as

From oblivion to glory for fascist will.

However, my correspondent Sgr. Rinaldi suggests that a better translation would be:

From oblivion to glory thanks to fascist will.

Sgr. Rinaldi also speculates that perhaps this inscription was the reason why, one night in 1945, a person or persons unknown blew up one of the Giovanni statues with gunpowder.

The fascist authorities placed Leonardo in a prominent position, in the front of the Logge di Bianchi, an elegant square at the southern end of the Ponte di Mezzo, the bridge that crosses the river Arno right in the center of the town. The name best translates as Central Bridge, but some sources give it as Means Bridge. While one translation of the Italian word mezzo is indeed "mean," that seems inaccurate in this context. The "mezzo" is the center of a city, and this bridge was the main one connecting the two parts of the center of medieval Pisa.

A bust of the minor nineteenth-century politician Felice Cavallotti was removed to make room for the square's new occupant, and the square itself was renamed Piazza XX Settembre (20th September Square). The name 20th September commemorates the date in 1870 when the city of Rome, until then under the rule of the Popes, was conquered by the new Italian state that had been founded in 1861. An irony presumably lost on the fascist authorities who renamed the square and moved Leonardo there, is that the official name they chose is written using Roman numerals, XX, not the Hindu-Arabic ones whose introduction to Italy and the west made him famous.

The statue continued to greet people crossing the Ponte di Mezzo until the end of the Second World War, when American and German troops fought an intense, month-long battle from opposite sides of the Arno. During the struggle, the bridge was destroyed, the piazza was badly damaged, and several surrounding buildings were either destroyed or left in partial ruins. Somehow, though, Leonardo miraculously survived with just the minor damage to his hands I noticed when I first saw the statue. Sgr. Rinaldi sent me a copy of a faded, but remarkable photograph showing Leonardo standing almost totally unscathed amidst a sea of rubble.

When the war was over, the statue of Leonardo was removed to allow for the rebuilding of the bridge and the square. Since the Camposanto itself was partly ruined and under reconstruction, it was not possible to return the statue to its original home. Instead, it was placed inside a city warehouse, where it was left forgotten for several years.

In 1966, the statue was finally taken out of storage and put on display in the Giardino Scotto. Unfortunately for poor Leonardo, the Giardino is a popular visiting spot not only for people but

FIGURE 5. The statue of Leonardo in 1945, standing in front of the right-hand arched entrance to the one remaining building, survived the battle that destroyed the Central Bridge in front of it, along with all the surrounding buildings. Photographer unknown.

also for birds, and over the years the statue not only became badly discolored by the riverside weather but also covered in bird droppings. The photograph of the statue I had seen several times on the Internet was an old one taken in the Giardino Scotto when the statue was dark and dirty.

Finally, in the years around 1990, the statue was removed and restored, and eventually placed back in the Camposanto where it began.

CHAPTER 5

A Walk along the Pisan Riverbank

PISA DAY 2. Evidence of Pisa's origins stretch back to almost 2,000 BCE, when it served as a transit port for Greek and Phoenician trade to and from Gaul. Later, the Romans also used it as a port. But it was not until another thousand years had passed that Pisa began to rise to the prominence it enjoyed when Leonardo was born.

In pre-Christian times, the Arno River that today divides the city into north and south opened up to a large lagoon just to the east, providing a natural port. The Romans called it the "Sinus Pisanus," although they were not the first to berth ships here. This was the flooded area I had crossed by train when I first arrived.

By Leonardo's time the lagoon had silted up and Pisa's status as a major shipping port alongside Genoa and Venice was sustained purely by the expertise and connections of its citizens, not its location. Indeed, sometimes, in dry weather, the Arno became too shallow for the larger ships to reach the city. Broad-beamed sailing ships and seagoing barges could generally get through, but the bigger vessels had to berth at Porto Pisano (nowadays part of the busy Mediterranean port city of

Livorno), several miles to the south of the Arno's mouth, along the seacoast. Their cargoes were then unloaded and carried into Pisa on narrow, oared galleys, or on river flatboats propelled by hand using a pole.

Other changes were also affecting the lives of the Pisans at that time. During the tenth century, as the 500 years of cultural stagnation known as the Dark Ages came to an end, European society began to develop and prosper once again. New farming techniques were introduced, the population started to expand, and national and international commerce began to develop. With few roads available, and most of those of poor quality, trading was carried out largely by river and sea transport. As a result, the bulk of western civilization was clustered around the shores of the Mediterranean.

From the tenth century onward, Pisa began to spread beyond its ancient Roman walls, with towers rising to the east and west, and to the south across the Arno. By the second half of the twelfth century, when Leonardo was growing up, a new, heavily fortified city wall was being constructed, to protect the city from attack both by Muslims—this was the time of the Crusades—and by rival Italian cities. At that time, cities often attacked one another as part of an ongoing political struggle between the Pope and Emperor Frederick II of the Holy Roman Empire.

Today, each bank of the Arno carries a major road, bustling with traffic—one of those roads being the Lungarno Fibonacci for part of its length. Lining those roads are row upon row of tall buildings, mostly residential, the walls of each one washed a characteristic yellow, orange, or brown, their windows flanked by wooden shutters painted green or dark brown. Five bridges span the river within the city confines, each joining one bank to the other.

On the second of my two days in Pisa, besides strolling around the old city and looking at the very buildings that would have

FIGURE 6. The Pisa riverbank, photographed in 2003.

surrounded Leonardo, I wanted to walk the length of the Arno between the city's two ancient walls. My earlier reading had led me to suspect that it was along the river bank that Leonardo, as a young child, was first exposed to the enormous power of arithmetic.

As the Arno's muddy brown water flows slowly and majestically toward the sea, the only vessels to be seen on its surface today are an occasional skull or rowboat. Eight hundred years ago, things would have looked very different. I tried to picture the scene Leonardo would have encountered had he walked along the same path—as well he might have.

Certainly, he would not have looked much like me. I was dressed in a sports shirt, blue jeans, and sneakers, but Leonardo would have looked far more colorful. Both men and women of that time wore bright colors: scarlet and blue, or green and yellow. Normal clothing for a young man from a prosperous family typically comprised linen undergarments, tight-fitting hose, a linen or wool tunic reaching down to just above the knees, a short, fur-lined coat, and soft leather shoes. In cold weather, Leonardo would also have worn an overcoat wrapped around his body and fastened at the right shoulder with a buckle, together with a hood or soft cap.[1]

I started my walk at the western end of the city. Had I set out eight hundred years earlier, the first thing I would have seen would have been a customs house. Pisa had two, one at each end of the city. This one, being nearest to the sea, handled vessels arriving from abroad. A typical incoming cargo might consist of sacks of grain from other parts of Italy, salt from Sardinia, bales of squirrel skins from Sicily, goatskins from North Africa, or ermine from Hungary. Some vessels had large doors in their sterns, which could be opened to allow horses from Provence to be led ashore. Particularly valuable imports were alum, destined for Pisa's leather industry, dyes for the textile manufacturers of Italy and northwest Europe, and spices from the Far East. Sacks of grain and barrels of wine were transferred from ship to barge for the voyage farther up the Arno to Florence. When all the cargo had been taken ashore, Pisan workers would reload the ships with goods for export: barrels of Tuscan wine and oil, bales of hemp and flax, and bars of smelted iron and silver.

Next to the customs house I would have seen the shipyard. Shipbuilding was a booming industry in twelfth-century Pisa, and her skilled craftsmen built ships not just for Italian clients but for France and North Africa as well.

Specially felled timbers were brought in by barge from the wooded uplands and unloaded by giant cranes. A fir trunk destined to become a mainmast could be 60 feet or more in length. The timbers were cut into planks in a large pit using a pit saw. Two men operated the saw, one standing on the ground above, the other down in the pit. They pushed and pulled the huge vertical blade, slicing through the log, as others shoved it lengthwise against the saw. The timbers

---

[1]   Much of my description of twelfth- and thirteenth-century Pisa in this chapter is taken from Gies and Gies (1969).

FIGURE 7. The Pisa Customs House, photographed in 2003, largely unchanged since Leonardo's day.

were shaped using heavy hammer-like adzes with curved iron blades. Despite the crude nature of their tools, skilled craftsmen were able to fashion the ships' timbers with remarkable precision.

The ships were assembled next to the water's edge. First the U-shaped frame pieces, or ribs, were fixed to a heavy keel to create a skeleton of the ship's hull. Then planks were nailed to the ribs, edge to edge, to form the ship's side. By fashioning the timbers accurately, the skilled Italian shipwrights were able to avoid having to overlap the planks, as was common with most other shipbuilders at that time. To make the ship seaworthy, caulkers worked their way over the entire hull, sealing holes and cracks with hot pitch.

Had I taken my riverside walk in the twelfth century, I might also have witnessed bars of crude iron being unloaded. Local ironmasters would have spent the winter at the iron mines on the islands of Elba and Giglio, supervising the mining and smelting of ore in furnaces at the site, before it was loaded onto ships and brought to Pisa, where skilled craftsmen would fashion it into tools, weapons, and armor.

At the water's edge by the Piazza San Niccola, and across the river in the Kinsica quarter, I would have seen—and smelled—tanners at work. They took raw hides shipped in from North Africa, and scraped them over a section of a tree trunk to remove hair and flesh. Then they soaked them in cold water and myrtle—the source of the distinctive smell—rubbing and beating them every day for up to six months, gradually transforming the raw skins into fine leather, ready to be cut and sewn into hats, belts, trousers, and other garments.

I might also have seen bales of wool being unloaded. In Fibonacci's lifetime, wool was just starting to replace leather for clothing. People who lived in the countryside had always spun and woven wool, and fulling (treating the woven cloth for softness and resilience) and dyeing were country industries, as was the sale of woolen cloth. During the early thirteenth century, these industries began to shift to the city. First tailoring moved in, then the sale of cloth, then the finishing processes, and finally, spinning and weaving made the move to the city.

Throughout my walk, scattered along the riverbanks among the shipbuilders, the berthed ships, and the tanners, I would have seen people bustling between dozens of colorful tents and improvised huts. These temporary places of business were erected by foreign merchants—Turks, Arabs, Libyans, and others—to display silks, carpets, vases, and other wares for sale.

At the end of my walk I would have come across the second customs house. This one, facing inland, served traffic from upriver. Shallow-draft boats and barges brought farm produce in from the countryside, or goods from Florence and other inland towns, to sell in Pisa's year-round market. And just beyond the customs house

I would have seen the Long Ford, where the Arno broadened and grew shallow enough to ride a horse across in low water.

The mathematician in me would also have seen something else in abundance along the twelfth-century Pisan waterfront: Numbers—lots of numbers. Wherever I looked, in the customs houses and at the wharves, I would have seen men using numbers: merchants measuring out their wares and negotiating prices, customs officers calculating taxes to be levied on imports, scribes and stewards preparing ships' manifests. I would have seen prices being recorded in *librae* (pounds), *solidi* (shillings), and *denarii* (pennies), with twelve *denarii* equal to one *solidus* and twenty *solidi* equal to one *libra*. I would have watched the scribes recording the values in long columns, using Roman numerals. They would have put their writing implements to one side and used a physical abacus to perform the additions, then picking up pen and parchment once again to enter the subtotals from each page on a final page at the end.

I imagined Leonardo taking a similar stroll along the riverbank as a child. He would have seen for real all the sites I saw merely in my imagination. Did he too see the numbers? Did he, a merchant's son, recognize the importance of those abstractions to the prosperity of his home city? And did he, even then as a young man, wonder if there was not a more efficient way of doing the math than to use those cumbersome Roman numerals, which required constantly going back and forth from pen and parchment to an abacus? Even if he had not, it is hard to imagine that memories of what my predecessor saw in his youth did not, in some way, prepare him for the momentous observation he would make in later life.

Admittedly, any other child growing up in twelfth-ventury Pisa would have observed the same scenes and been exposed to the same influences. And no doubt some of them, in adult life, also would have crossed the Mediterranean and seen Arab merchants using the Hindu number system, just as Leonardo did. His genius lay in recognizing that the introduction of a more efficient way of handling numbers would quite literally revolutionize world trade—and with it the world. And he was shrewd enough to know exactly what to do to ensure that the new method would be widely accepted.

My time in Pisa had been very brief—just two days—but already I felt I was starting to get to know something about Leonardo. I resolved to return the following year. Before I did, I intended to do something I had hitherto been putting off. I was going to try to read *Liber abbaci*. Since the book had never been translated (I would not hear of Sigler's forthcoming book for two more days), that would mean refreshing my long-unused high school Latin. Though, truth be told, mathematical Latin is so narrowly focused and stylized, it is not too difficult for a mathematician to read, provided you have a Latin–English dictionary to hand.

What would I learn from reading Leonardo's own words? A conversation I had had in Bologna on my way from Trento to Pisa a few days earlier had given me some idea of what to expect.

CHAPTER 6

A Very Boring Book?

"You know, it [Liber abbaci] is a very boring book. It's just one problem after another. Here is a problem, here is how you solve it. Here is a problem, and this is how you solve it. Then another problem and its solution. Then another. And another."

The speaker was Giulo Cesare Barozzi, professor of Mathematics and Engineering at the University of Bologna. He was tall, heavyset with broad shoulders, seemingly in his late sixties, with silver hair and a small, well-trimmed beard, and the distinguished demeanor of a highly cultured, European scholar.

I had broken up journey from Trento to Pisa in the ancient and extremely attractive university city of Bologna to give a lecture on the theory of mathematical cognition I had presented in my book, *The Math Gene* (2000). The Italian translation had become a minor best seller in Italy and so for a while I was enjoying a small measure of "crowd pulling" ability. My lecture drew an audience of about 60 Italian high school mathematics teachers, who had traveled from towns and villages as far away as Venice.

FIGURE 8. Professor Giulio Cesare Barozzi of the University of Bologna. Photograph courtesy of Professor Barozzi.

Professor Barozzi had introduced me to the audience and moderated the discussion that followed my talk. But speculating on how the human brain had acquired the ability to do mathematics was not what I wanted to talk to Barozzi about. With my speaking commitments in Italy coming to an end, I was starting to think about the visit to Pisa I was about to make, and my curiosity about one particular thirteenth-century mathematician who had lived there.

It was a delightful spring day, and the two of us were sitting at a tiny, round, cloth-covered table at a typical outdoor cafe in the Via Monte Grappa, a narrow cobble-stoned street in the heart of Bologna. Every few moments, a noisy automobile or motor scooter sped past, smelling of gasoline and belching thick, black exhaust, interrupting our conversation.

The sun—still warm in the late afternoon—cast shadows on the soft-colored beige and pink buildings around us. Waiting for our drinks to be served, we nibbled on potato chips as I told Barozzi about my fascination with Fibonacci, and my intention to write a book about him. (I started to call him Leonardo—his real name—when, through my research, I got to know him a little better.) He was surprised at my interest, and felt sure my search would prove fruitless. "Nothing is known about him," he declared. "The few things you read are all made up."

Some of the the things Barozzi told me were already familiar to me from my initial research before my trip, but others were new. Still other parts of the story remained, waiting for me to learn, as I delved further into my research. On occasion I would discover that Barozzi too had been misinformed.

He began with a brief outline of the basic history of how modern arithmetic came to Europe.

"The highly efficient number system we use today, where just ten basic symbols 0, 1, 2, 3, 4, 5, 6, 7, 8, 9 and a decimal point suffice to express any (positive) number, was developed in India during the first seven centuries after Christ, along with the now-familiar rules for performing basic arithmetic with them."

He paused to allow a small but very loud motor scooter roar past.

"Arabic-speaking traders who traveled back and forth along the famous Silk Road to India learned the Hindus' system and used it in their commercial transactions, spreading the new knowledge to the Middle East and North Africa. Arabic scholars wrote books explaining how to use the system, and increased its power by developing what we now recognize as the beginnings of algebra, the very word *algebra* being of Arabic origin."

"Leonardo's *Liber abbaci* was the book primarily responsible for bringing the Hindu-Arabic system, as we now call it, from the shores of North Africa across the Mediterranean to Western Europe," he continued.

"The university library in Bologna had a copy of *Liber abbaci*. It was a printed book dating from the nineteenth century, not an old or particularly valuable manuscript copy." I found this disappointing, given the fact that the University of Bologna was founded in Fibonacci's time, and was the first ever university in Italy—and most likely in all of Europe. Indeed, it was Bologna that had given the world the word "university."

The drinks arrived. Barozzi had ordered a small glass of freshly squeezed lemonade, but after lecturing for 90 minutes I had worked up a thirst for a very large, and very cold, beer.

"The best copy is in the Vatican Library in Rome," my host continued. "The first edition came out in 1202. I don't think any copies of that version exist today. Fibonacci wrote a revised edition a few years later. All manuscripts were written by hand in those days, of course. The first printed edition was published by Boncompagni, the publisher in Rome, sometime in the nineteenth century."

I had read earlier that Leonardo had in fact completed his second edition in 1228, and that it carried a preface stating that "new material has been added from which superfluous had been removed" [*sic*]. The printed edition prepared and published by Baron Baldassarre Boncompagni—an Italian bibliophile and medieval mathematics historian—that Barozzi had referred to was published in 1857, and was based on a late-thirteenth- or early fourteenth-century copy of Leonardo's 1228 version. It was the first of a two-volume, printed collection of all of Leonardo's works that Boncompagni published in Rome under the title *Scritti di Leonardo Pisano*. The second volume, containing all of Leonardo's other works, appeared in 1862.

According to a medieval scholar I consulted in Siena after my visit to Pisa, the best existing manuscript, and possibly the oldest—written no later than just a few years after Leonardo's death—was actually in Siena, and not in Rome, as Barozzi had declared. By his own admission, my gracious host in Bologna was not a Fibonacci scholar.

Barozzi's lack of knowledge about Italy's most influential mathematician of all time was typical of almost every Italian mathematician I have spoken to. Indeed, most knew far less about their illustrious countryman than did Barozzi. In Siena a few days later I would speak with a prominent mathematician who responded to my expression of interest in Fibonacci by asking, "Wasn't Fibonacci the man who developed the decimal number system?" Most mathematicians are interested in mathematical results, not the people who discover them.

The lack of concern mathematicians have for the people who make the key discoveries in the discipline—the ones who solve the problems and prove the theorems—reflects the nature of the subject. Although mathematics is created by people, mathematical truth is completely independent of human judgment. Mathematical knowledge, once discovered, is immutable and eternal. (Though the way those mathematical truths are perceived and expressed can vary from one culture to another, and over time.) In all other areas of science, in contrast, new discoveries replace older ones. In astronomy, for example, the scientific theories of the ancient Greeks had to be abandoned when Newton put forward his ideas of gravitation, and in due course Einstein's theory partially overturned that of Newton. Not so in mathematics. For example, once Euclid had proved, around 350 BCE, that there are infinitely many prime numbers, that was the end of the matter. You could take Euclid's classic book *Elements* and use it as a textbook for a geometry class in any high school today. In fact, until the 1960s, schools all around the world did just that.

For many mathematicians, part of the attraction of mathematics is that it forms its own abstract world, quite separate from (although originating in and generally of great application to) the real world of our everyday experiences. The lives and personalities of the individuals who develop new methods and make the discoveries play no role in the mathematics itself, and hence are generally dismissed (by mathematicians) as being of little or no interest.

The nature of mathematical creativity is such that if person A had not come along to make some new mathematical discovery, then person B would have made the same discovery soon after, and if not B then C.

The only exception I know to this are the results of the early twentieth-century Indian mathematician Srinivasa Ramanujan. Had he not lived, it is likely that no one would have discovered many of the things he did. But his case is extremely unusual in several respects, as was the mathematics he produced. In fact, it is often the case that two or more people make the same discovery independently, at the same time. With the people involved seemingly so unimportant, why bother trying to find out about A or B (or C)?

The timeless nature of mathematics also means that, for its practitioners, historical details are of little relevance. To the mathematician, it doesn't matter *when* someone first proved Pythagoras's theorem, or even why; the important thing is that we know it's true. Even most scholars who work in the academic discipline called "the history of mathematics" are in general not particularly interested in the *people* who did the mathematics. Their focus is on the development of the *ideas*, and how one train of thought led to another. That, after all, is what mathematics is about. Its very essence, and the source of its immense power, is that it abstracts *away from* the real world.

This perhaps explains why I was not surprised to meet Italian mathematicians who knew virtually nothing about their predecessor, who has had more influence on the course of the world than almost any other mathematician in history. Professor Barozzi's knowledge of Leonardo was actually far greater than most. Like the mathematician I talked with briefly in Siena a few days later, the majority of my colleagues around the world probably have only the sketchiest knowledge of how numbers found their way into our everyday lives.

One of the difficulties in trying to appreciate the significance of *Liber abbaci* is that, in today's world, numbers are so very commonplace. We simply take them for granted. Sometimes, when I meet people who say they were not good at elementary arithmetic in school, I ask why they did not do better. The most frequent answer I get is that they found it boring. (You have to get to fractions or algebra before people start to answer that they could never understand it.)

Yet the Hindu decimal number system and arithmetical methods for calculating with them that Leonardo learned from the Arabic-speaking traders and scholars in North Africa are two of mankind's greatest and most revolutionary inventions. Why do we marvel at the pyramids as a feat of human creativity and overlook the far more inventive mathematical ideas that made their construction possible?

The answer, surely, is that the pyramids are visually dramatic "one offs" that could easily not have been built, whereas numbers and arithmetic are a fundamental part of the very fabric of our lives. We can no more imagine a world without numbers and arithmetic than we can one without words and sentences.

We rejoice in Shakespeare but groan at the idea of diagramming sentences. Yet both language and numbers are human creations. In the case of language, all the evidence points to the use of individual words or perhaps small groups of words to communicate basic needs or instructions (for example, "Me hungry") going back perhaps as many as three million years. Full-blown language, on the other hand, with its complex rules for constructing sentences that allow us to talk about anything we want, seems to be much more recent, somewhere between 75,000 and 200,000 years.

Numbers are, to the best of our knowledge, even more recent. Just as words came long before sentences, counting came before numbers. Notched bones that appear to have been used for counting have been found that date back around 35,000 years; clay tablets bearing symbolic marks for actual numbers, less than 10,000 years.

But, while both language and arithmetic were passed on by inscriptions on tablets, those tablets were not "handed down" from a deity; they were human creations—the products of the human mind. It is perhaps inevitable, though to my mind a little sad, that the creations that turn out to be the most profound for our lives eventually become so commonplace that we no longer see them for the huge accomplishments they are.

Our drinks were almost finished, and a large beer following a two-hour lecture session with 60 high school students had given me an appetite. So when Barozzi told me he had to leave, I was ready to say goodbye. Though I already knew most of what he had recounted about Leonardo and *Liber abbaci*, it somehow seemed more real to hear it directly from an Italian scholar. Only later would I discover that, while broadly true, his account glossed over several complexities, which I would describe in the Leonardo book I eventually wrote, *The Man of Numbers*. A much more complete, and accurate, account would come from the historian of medieval mathematics I was to meet in Siena a few days later. Before that, I was scheduled to spend a couple of days in Pisa, in search of Fibonacci monuments and memorabilia. I hoped the good weather would continue.

CHAPTER 7

Franci

With my visit to Pisa behind me, I had just one more stop to make before leaving Italy. Upon hearing of my trip to Italy early in 2002, my old friend Franco Montagna, a professor of Mathematics at the University of Siena, had invited me to give a lecture in the Mathematics Department.

I was looking forward to seeing him again after several years during which our paths had not crossed. Back in 1984, Montagna had arranged a three-week stay for me as a visiting professor at the University of Siena. As it happens, my popular mathematics writing had already been part of my life for more than a year when I took up temporary residence in a gorgeous, university-owned, old monastery (Certosa di Pontigano) just outside of Siena, but it was still very much a minor part. My main interest continued to be mathematics research, not mathematics outreach, and it was my research work that led to Montagna's invitation.

I had met Montagna a few times since that 1984 stay, once in Siena, the other times at conferences in various locations around the world. Over the years, as my research interests grew further away from mainstream mathematics, there was less motivation for Montagna and I to interact, and as a consequence our paths crossed much less frequently.

I had fallen in love with Siena and the rolling Tuscan hills that surround it, but for many years I did not visit. So it was with considerable excitement that I boarded the train for Siena, where I would meet up with my old friend once again. We had a lot to catch up on—and I would be back in Siena once again.

Sightseeing and catching up with an old friend were not the only things I wanted to do, however. In our exchange of email messages setting up my visit, I had told Montagna of my desire to find out what I could about Leonardo of Pisa, and he had suggested that he introduce me to a mathematical colleague of his at the University: Professor Rafaella Franci, the director for the University's Center for the Study of Medieval Mathematics.

Montagna's area of specialization in mathematics was Logic, the overarching subject I had pursued for my PhD and the focus of my visit to Siena two decades earlier. Neither of us knew much about the history of mathematics. Montagna was the Siena mathematician who, when we started talking soon after my arrival, asked if Fibonacci was the man who invented the decimal number system.

In contrast, his colleague Rafaella Franci, to whom I was to be introduced, had devoted much of her career to the study of medieval mathematics, particularly in Italy and Europe.

Though I did not know it at the time, Siena, far more than Pisa, is where I would be spending most of my Leonardo-project research time.

Like many Italian towns dating from medieval times and earlier, the city of Siena, about 40 miles south of Florence, is built on a hilltop. Two tall towers dominate the skyline. One tower is part of the clay-red, brick-built town hall, and looks down on the Piazza del Campo, the cobblestone main square of the city that has always been the center of life in Siena. Nowadays, the Campo is filled with tourists for much of the year. The younger ones sit on the ground and enjoy the afternoon sun, and perhaps feed the pigeons; the older ones stroll around or sit at one of the many outdoor cafes that ring the square, sipping a cappuccino or a glass of Chianti. In medieval times, the Campo was the focus of the city's political life.

The second tower, a short distance to the northwest, is part of the cathedral—the Duomo—whose distinctive, black and white, zebra-striped, marble facade can also be found, although in far less dramatic form, on the cathedrals of Pisa and Florence.

If you step out of the cathedral and head left across the Piazza del Duomo, you come to a narrow street called the Via del Capitano. Follow it for a few paces and there, on your right, you will find a white stone building with the Roman numeral XV on the wall. The building dates back to the seventh century, and was originally a palace. When I made my trip, it had a number of different inhabitants. One of them was the Mathematics Department of the University of Siena, which had been there since 1973. That was where Franci's institute was housed. The plan was for me to meet with Franci on my first day, then return to the department the following day to give a lecture on *The Math Gene*.

[Much to the sorrow of visitors like myself, but not the Siena mathematicians, the university subsequently moved the Mathematics Department out of that delightful building to another location, which is close by. Though not as beautiful to look at, the new facility provides the faculty with a far more spacious and comfortable working environment—with some spectacular views over the Tuscan vineyards.]

The first day of my visit was hot and humid, and the Via del Capitano was filled with tourists making their way to and from the Duomo. I stepped through the doorway leading to Via del Capitano XV, and immediately found myself in a small,

FIGURE 9. The entrance to the University of Siena Mathematics Department, where it was located when I visited in 2003. Photograph taken in 2009, after the department had moved elsewhere.

enclosed courtyard that was strangely cool and quiet after the heat, noise, and bustle of the street outside. I must have spent many hours in that building in the spring of 1983, but 20 years of academic travels and visits had all but obliterated any memory of the place.

The mathematics department occupied the upper three floors of the building's four stories. To reach it, I had to climb the elegant staircase to the right of the yard. Montagna greeted me on my arrival and led me through the building to Franci's office, where he said she was waiting for me.

The section that housed her center was a contrasting mixture of the old, artistic, and expensive, and the new, functional, and parsimonious. In the large rooms that served as an entrance foyer and classrooms, elegant dark-wood-paneled walls and gilt-painted frescos on the ceilings harkened back to an earlier era. Yet in many of the adjacent rooms, cheap plywood walls had been erected to create tiny faculty offices.

One of those makeshift offices, tucked away at the rear of the building, was Franci's. [In contrast, Franci's spacious office in the new mathematics building is on the fourth floor, with a spectacular view from the highest point of the hilltop city of Siena across the rolling Tuscan hills to the south.] She stepped out of the door to greet me, a short, gray-haired woman with wire-framed glasses. She was wearing a purple cotton trouser suit and a woolen jumper whose color matched her hair. Since there was not enough space to entertain visitors, she led me to another part of the building and ushered me into a dusty and dingy conference room, surrounded by shelves crammed full of fading mathematics journals. An old, abandoned microfiche viewer stood in one corner. We were joined a few minutes later by Professor Paolo Pagli, one of Franci's colleagues in the Center for the Study of Medieval Mathematics.

Though I wanted to learn as much about Leonardo as I could from Franci, one particular question was foremost in my mind. Why was *Liber abbaci* the book that changed the world? There was, after all, another natural candidate, a book that had been written more than 350 years earlier.

In the eighth century, under the `Abbasid dynasty, the Islamic Empire, barely a century old, went through a remarkable period of cultural and intellectual growth. Caliph `Abbasid al-Mansur, who reigned from 754 to 775, founded the city of Baghdad on the banks of the Tigris, making it the capital of the empire. Its location made it a natural crossroads, the place where East and West could meet, and the new city quickly became a major cultural center and a center of learning. In Baghdad, Arab scholars studied and translated Greek and Hindu mathematical texts, including the Indians' place-value system for writing numbers and for doing arithmetic.

FIGURE 10. Rafaella Franci in her (new) office at the University of Siena in 2009.

Among those scholars was a distinguished ninth-century mathematician called Abū ꜥAbdallāh Muḥammad ibn Mūsā al-Khwārizmī,[1] who lived from ca. 780 to ca. 850. It has been suggested that his name indicates he was from the town of Khwārizm (now Khiva), on the Amu Darya River, south of the Aral Sea in what is now Uzbekistan. (Khwārizm was part of the Silk Route.)

As I noted earlier, al-Khwārizmī is best known today for having given the world of mathematics two important names: *algorithm* and *algebra.* He wrote several books, including one about the Hindu number system: *Kitab al-hisab al-Hindi* [Book of Hindu reckoning].

---

[1] Not Abū ꜥJa'far Muḥammad ibn Mūsā al-Khwārizmī, as given incorrectly in some sources.

Only Latin translations (bearing the name *Algoritmi de numero Indorum*) of that work have survived to this day. Many of those translations began with the phrase "dixit Algorismi" [so says al-Khwārizmī], a practice that led to the adoption in medieval times of the term *algorism* to refer to the process of computing with the Hindu numerals. Hence our modern word "algorithm."

The word "algebra" comes from another of al-Khwārizmī's manuscripts: *Kitab al jabr w'al-muqabala*, which translates roughly as "restoration and compensation." The book is essentially an algebra text. It begins with a treatment of numbers, followed by a discussion of quadratic equations, continues with some practical geometry, goes on to discuss simple linear equations, and ends with a long section on how to apply mathematics to solve inheritance problems. After al-Khwārizmī, algebra became an important part of Arabic mathematics. (Prior to al-Khwārizmī's work, algebra was largely a practice developed and used by traders.) Arabic mathematicians learned to manipulate polynomials, to solve certain algebraic equations, and more. (All of this was done without the use of symbolic notation.) It was the Arabic phrase *al jabr* in the book's title that gave rise to our word "algebra."

Over the ensuing decades, a number of other scholars carried forward al-Khwārizmī's work. In the eleventh century, a Persian geographer, historian, and physicist called al-Biruni, after traveling in India and learning Sanskrit, wrote a book about Hindu arithmetic, basing it on al-Khwārizmī's arithmetic text, to which he also supplied a commentary.

Another famous Arabic mathematician of that era who studied al-Khwārizmī's books was 'Umar al-Khayámmi, known in the West as Omar Khayyám, who lived approximately from 1048 to 1131. Although he is remembered in the West primarily as a poet, he was a well-known mathematician, scientist, and philosopher, who did major work in all those fields.

In the middle of the twelfth century, scholars translated both of al-Khwārizmī's books into Latin. The Englishman Adelard of Bath may have produced one translation of the arithmetic book, *Algoritmi de numero Indorum*. In 1145, Adelard's countryman Robert of Chester made the first of what were to be several translations of al-Khwārizmī's algebra book from Arabic into Latin. Those translations would, in due course, become a significant source for Europeans who wanted to learn the new mathematics.

Yet, despite this series of al-Khwārizmī-inspired books, including al-Khwārizmī's two originals, it would be Leonardo's *Liber abbaci*, not his Arab predecessor's much earlier books, that would earn the honors for bringing both Hindu numbers and algebra to the West. This was true even though Latin translations had been completed more than a half century earlier. Why was this so? That question had bothered me ever since I had determined to write a book about Leonardo.

Knowing that my visit was to learn about Leonardo, as soon as Montagna had introduced us, Franci began, without prompting, to tell me that she had been studying Leonardo's works and their influence for more than 20 years. "What would you like to know about him?" she asked. I jumped straight to my main question. "Why was *Liber abbaci* the book that popularized the Hindu number system and not the earlier arithmetic text by al-Khwārizmī, which had been translated into Latin in the mid-twelfth century, some 30–40 years before *Liber abbaci* appeared?"

"There are several reasons," Franci explained. "First, al-Khwārizmī's methods were different from the ones Leonardo later adopted, with Leonardo's clearly the superior. Al-Khwārizmī's approach involved cancellation and over-writing, which made it impossible to track the course of the calculation after it was completed.[2] So checking a calculation was not possible. Nor was going back to correct an error; you had to start all over again from the beginning. Leonardo used methods more like those taught today, where steps in a calculation are laid out one after another in an orderly fashion that spreads out over the page, making both checking and correction an easy matter."

Another difference, Franci continued, was the way the two books were distributed. Although al-Khwārizmī's text was written for use by merchants as well as scholars, in the West the Latin translations were read only by learned men. In contrast, the methods described in *Liber abbaci* were, from the outset, mastered by the merchants and other businessmen.

Though the dense, scholastic Latin and the detailed Euclidean-style proofs of Euclid found in *Liber abbaci* may well have prevented all but a few scholars from reading it, Franci explained, Leonardo's mammoth text was the inspiration for what would be many shorter books on practical mathematics—often little more than extracts from Leonardo's work—written in vernacular Italian and used by teachers in the "abbacus schools." These public schools began in the mid-thirteenth century, and were where the merchants sent their sons.

"*Liber abbaci* is not really a 'how-to book,'" Franci cautioned. "It's definitely a mathematics book. Leonardo develops everything with mathematical rigor, and explains the mathematical reasoning. What made it valuable to the merchants are the examples. There are hundreds of examples of practical calculations of the kind merchants needed to carry out every day. A merchant might not

---

[2]   I learned later that al-Khwārizmī's algorithms were developed to perform calculations on a dust board, where erasing is easy. The Arabic-speaking mathematicians subsequently developed different procedures for use with pen and paper, and these are the ones Leonardo described.

be interested in why a particular method works, or may not have the background to follow it, but he can still follow the examples and learn how to do it."

Still another factor that led to *Liber abbaci* being far more influential than al-Khwārizmī's writings, my host continued, was that the medieval Italian universities—which did use (Latin translations of) al-Khwārizmī's books—were not very good, and their mathematics instruction was restricted to some study of algorithms and bits of geometry. Overall, they did not have much impact on ordinary life at the time. The merchant classes who read the many, shorter texts *Liber abbaci* gave rise to, in contrast, were a powerful and influential group.

"Was Leonardo familiar with al-Khwārizmī's work?" I asked Franci. "Absolutely," she replied at once. "Al-Khwārizmī's books and commentaries on them were well known in the whole Arab world, where Leonardo traveled and studied, as he acknowledges in the introduction to *Liber abbaci*."

"What we do not know," she continued, "is what other Arabic sources he used for his book. There have been comparative studies of his other books on algebra with Arabic works that have yielded information about some of his sources for those other texts. But no one has carried out a similar investigation for the first part of *Liber abbaci* [the part that describes the Hindu-Arabic number system]."

When I asked why, Pagli answered. "Because there is just too much material," he said simply.

Franci stressed that Fibonacci made no claims of originality in *Liber abbaci*, although he did so in another of his books, *Liber quadratorum*. I sensed that she had a fondness for Leonardo every bit as great as mine, and like me she did not want anyone to form a false impression of the man or his intentions.

Both Franci and Pagli were not only intrigued at the project I had set out to complete, they were also highly skeptical that I would get very far. They both repeated what I had already learned from others, that there is almost nothing known about Leonardo's life. That, of course, would prevent me or anyone else from writing a standard biography of his life, and explained why none had ever been written. But that was not my aim. I did not want to write a historical biography in the familiar sense. I wanted to get, and then convey to others, a sense of the man who brought numbers to the West. And I wanted to do so in part by understanding the time when he lived and in part through his mathematical writing.

As things were to turn out, I would gain much of that understanding when I realized that, by a fortunate accident of history, those of us alive today have experienced a revolution very similar to the one occasioned by *Liber abbaci.* Indeed, the parallels are uncanny. But that is getting ahead of my story.

I had two more questions I wanted Franci to answer. The first was, how many manuscript copies of *Liber abbaci* had survived from Leonardo's lifetime, and where were they now?

The bad news, Franci confirmed for me, was that no copies of Leonardo's original 1202 edition have survived. Of his 1228 revision, 14 copies still exist, she said. Seven of them are substantial and seven mere fragments, consisting of between one-and-a-half and three of the book's fifteen chapters. Of the seven reasonably good copies, three are complete or almost complete and are generally regarded as the most significant. All three are in Italy.

As Professor Barozzi correctly observed when we met in Bolgona, one good manuscript is in the Vatican Library in Rome, where it bears the reference mark Vatican Palatino #1343. This manuscript, from which chapter 10 is missing, is believed to date back to the late thirteenth century. Another, also believed to date from the late thirteenth or perhaps the early fourteenth century, is in the Biblioteca Nazionale Centrale di Firenze (BNCF—Florence National Central Library), where it is listed in the catalogue as Conventi

Sopressi C.1.2616. This manuscript is complete, which probably explains why the publisher Baldassarre Boncompagni used it as the basis for his first printed edition in the mid-nineteenth century, even though it is not the best preserved, nor perhaps the oldest. The third good manuscript, generally believed to date from the thirteenth century as well, and according to some scholars possibly the oldest, is housed in the Biblioteca Communale di Siena (Siena Public Library). This is the manuscript Franci used for her own research.

I learned subsequently that of the remaining manuscripts, four are housed in the BNCF, along with the one I mentioned above; one is in the Biblioteca Laurentiana Gadd in Florence (Gadd. Reliqui 36, dated to the fourteenth century); one is in the Biblioteca Riccardiana in Florence; one in the Biblioteca Ambrosiana in Milan; one in the Biblioteca Nazionale Centrale in Naples; and three in Paris (one in the Bibliothèque Mazarine, two in the Bibliothèque National de France).[3]

I resolved that on my next visit to Italy, possibly as early as the following year, I would seek out all three good manuscripts. My mostly forgotten school Latin might not be adequate for me to read everything, but I should be able to follow the symbolic mathematics, and reading pages that possibly dated back to Leonardo's lifetime would help me create in my mind a sufficiently good sense of the author to be able to write a book about him.

My final question for Franci, before we left the Department and headed out to the Campo for lunch, was: Why had 800 years passed without anyone translating *Liber abbaci* from Latin into a modern language?

Pagli explained why no one had taken the time to produce a translation. "There is just too much material. The mathematics

---

[3] For more details, see Hughes, 2004.

itself was available in many other, much shorter books written by others, and many of those books were translated into several languages. The only people who would be interested in reading Leonardo's original were scholars, and they could read Latin."

"But there will soon be an English translation," Franci added. "An American is working on one. I believe it will be published later this year."

By evening I could not recall what I ate for lunch after our meeting. I had been too excited at the possibility of being able to sit down and read an English translation of Leonardo's actual words. No need to struggle through the original Latin with a Latin-to-English dictionary to one side as I had expected. The moment I returned home to California, I would order a copy of the soon to be published English version.

CHAPTER 8

Publishing Fibonacci: From the Cloister to Amazon.com

The first copy of Liber abbaci appeared in Pisa in 1202, but no one knows where or when Leonardo actually wrote it. Its exact origin is yet another unknown factor of Leonardo's life.

When Leonardo finished writing his work, he would have taken the manuscript to a local monastery to have copies made by the monks. This was a laborious method of publication. It could have taken a year or more to copy a manuscript as long as *Liber abbaci*, though we have no idea how long that first, 1202 edition was.

Hand copying created major problems for the historian, not only because of copying errors but also—and this was an even greater problem—because scribes were not averse to making their own additions or deletions, and even copying material from

The early part of this chapter contains some overlap with sections of *The Man of Numbers*, but only enough to make the story told here self-contained. I do, however, provide some additional information, based on present-day analyses, of some of the contents of *Liber abbaci* that I did not include in my earlier book.

another manuscript, often without leaving any indication that they had done so.

We do not know where Leonardo was living when he wrote the book, nor do we have any way of knowing what its contents were when it first appeared. After the first edition had been published, he kept making changes and adding to its contents, culminating in a second edition that he published in 1228. Only this second edition has survived to the present day. Franci has speculated[1] that the first edition focused only, or largely, on arithmetic, and that Leonardo added much or all of the extensive treatment of algebraic methods in the 1228 edition.

From brief remarks Leonardo put in the introduction to the 1228 edition, we know that, after spending some time in Bugia, he traveled extensively around the Mediterranean, talking with mathematicians and learning more about Greek and Arab scholarship in general. In particular, he visited Egypt and Syria in North Africa and Greece, Sicily, and Provence in Europe. He also spent several years in Constantinople.

Given *Liber abbaci*'s enormous length, and its huge number of carefully worked out examples, my own guess as an author of mathematics textbooks is that Leonardo probably started working on it soon after he left Bugia, where he first encountered Hindu-Arabic arithmetic and recognized its huge significance. Most likely he carried the manuscript with him on his travels, working on it and discussing its contents with the mathematicians he encountered at each location.

As I had learned from my reading prior to my trip, and had discussed with Professor Barozzi in Bologna, the earliest complete *printed* copy of *Liber abbaci* (the 1228 edition) was printed by Baldassarre Boncompagni in Rome in the period 1857–1862. Boncompagni

---

[1] Private conversation, 2009.

based his edition on the one with the reference number *Conventi Soppressi* [2] C.1. n, dated to the late thirteenth century, now located in the Biblioteca Nazionale Centrale in Florence.

The Boncompagni edition, comprising 387 densely packed pages, formed the basis of the 2002 translation of *Liber abbaci* into English.

A quick online search upon my return home from Italy revealed that the American translator Franci had told me about was Laurence Sigler, and the publisher was the German academic book publisher Springer-Verlag. That was lucky. Springer-Verlag had published several of my books, including the first two I ever wrote, and I knew several of the senior editorial staff. So if I had questions about the publication side, I could likely find answers.

I recognized Sigler's name from his earlier translation of Leonardo's (much shorter) algebra book *Liber quadratorum* [Book of squares], which I had already obtained and read.

I placed an advance order for the book on Amazon. It was the first time in my life that I'd ordered a book prior to publication. The fact that the English-speaking world had been waiting for a translation for 800 years did not allay my impatience to read Leonardo's words for myself; I wanted a copy at the first possible opportunity.

Around the time I placed my order, a news article observed that Amazon was receiving tens of thousands of advance orders for the next volume in the *Harry Potter* children's fantasy book series. I suspected that mine was the only advance order they got for Leonardo's master work. (I would later discover that a finance professor at Yale placed an order as well.)

When the bulky package from Amazon arrived, I was excited at the prospect of finally being able to read the words that so

---

[2] Loosely translated as "Monastery Press," but more literally as "suppressed convents" (for "suppressed convent manuscripts"), which reflects the history of these manuscripts.

changed the course of history. I hoped that, as a mathematician, I could gain insights into Leonardo by reading his written work, particularly a lengthy textbook that gives plenty of scope for the author's character to show through.

My first reaction was amazement at the sheer size of the book—more than 600 pages, set in a fairly small typeface. Of course, there are plenty of books of a similar size or even longer; some calculus textbooks these days have 1,000 or more pages. But *Liber abbaci* has a very narrow focus; its 600 pages are devoted almost entirely to what we today consider elementary arithmetic and algebra. Holding the volume in my hands gave me a physical—and very heavy—appreciation of the comment I had read that one of the strengths of Leonardo's book was its large number of practical examples.

Growing up in modern society, the beginning math student today does not require (and would not tolerate) 600 pages of (mainly) worked examples on basic arithmetic and algebra, but Leonardo's intended readers were growing up in a very different environment. As a result, examples are what occupy most of *Liber abbaci*'s 600 pages.

At first I tried to read the book through the eyes of a person living in the thirteenth century, who had not encountered the Hindu-Arabic number system before. But I found that impossible. Numbers, in particular numbers written in the Hindu-Arabic fashion, are simply too great a part of what we are today. We are surrounded by numbers, and we become familiar with them at a very early age. We simply cannot forget what we know about numbers.

Some indication of the degree to which we identify numbers with their Hindu-Arabic symbols is provided by the following experiment, which psychology students are often asked to perform. You present subjects with pairs of numbers on a computer screen and ask them to press one of two keys as quickly as you can to say which one is written in the bigger font. For instance, which of **3** and **7** is in the bigger font? Which of **4** or **8**? The computer times your response each time. The second of these two takes you longer. Even though you know you should ignore the number and just concentrate on the size of the symbol, you cannot. It is harder, and will take you measurably longer, to say which one is the larger font if the larger of the two numbers is in the smaller font than if the font sizes agree with the number sizes. Your mind simply cannot disassociate the number from the symbol.

For Leonardo's readers, the situation was very different. To most people at the time, symbolic numbers and their arithmetic were alien. Arithmetic was something practiced by the merchants, who used an abacus (in Leonardo's Italy it took the form of a ruled board on which the merchant moved pebbles) or finger reckoning (an intricate system that allowed computations with answers up to 10,000).

It is telling that Leonardo devoted more pages to explaining how to write numbers than how to add them. Leonardo spends *four* whole *pages* (in the densely packed Sigler book) explaining the meaning of a units column, a tens column, a hundreds column, and so on. To today's reader, this seems absurd, but it just confirms the degree to which modern society has incorporated the Hindu-Arabic place-value decimal number system into daily life.

When Leonardo promises to explain "how the numbers must be held in the hands," he means that quite literally. Medieval scholars and traders used a fairly standardized system of finger positions to represent different whole numbers, and manuscript copies of *Liber abbaci* (and other arithmetic books that were to follow) often included a drawing showing the various finger positions used to store different numbers. Today we use numerical symbols marked on a sheet of paper for the same purpose.

Although Western scholars had learned about the Hindu-Arabic number system as early as the tenth century, it was not until Leonardo's book came along, with its many pages of everyday, practical examples, that it became accessible first to the merchants and eventually to everyone.

Leonardo went to such great lengths to explain how to use the new number system, it seems likely that, for all its novelty, most readers would have been able to master it. But that is not to say they found the book easy to read. They surely did not. Leonardo introduced many rules for computation and solving problems, and although he illustrated all of them with copious examples, some are very difficult.

Having explained the number system in his first chapter, Leonardo divided the remainder of the book into 14 additional chapters. I discussed them all in *The Man of Numbers*, and for convenience I provide a summary in the appendix to this book. More relevant to the present story is the material that Leonardo provides in a prologue to the text. It begins with a dedication:

> You, my Master Michael Scott,[3] most great philosopher, wrote to my Lord[4] about the book on numbers which some time ago I composed and transcribed to you;[5] whence complying with your criticism, your more subtle examining circumspection, to the honor of you and many others I with advantage corrected this work. In this rectification I added certain necessities, and I deleted certain superfluities. In it I presented a full instruction on numbers close to the method of the Indians, whose outstanding method I chose for this science. And because arithmetic science and geometric science

---

[3] Michael Scott was a philosopher in the court of Emperor Frederick II.
[4] Leonardo is referring to Frederick II.
[5] Sigler's translation is based on the second edition of the book, published in 1228.

are connected, and support one another, the full knowledge
of numbers cannot be presented without encountering some
geometry, or without seeing that operating in this way on
numbers is close to geometry; the method is full of many
proofs and demonstrations which are made with geometric
figures. And truly in another book that I composed on the
practice of geometry[6] I explained this and many other things
pertinent to geometry, each subject to appropriate proof.
To be sure, this book looks more to theory than to practice.[7]
Hence, whoever would wish to know well the practice of this
science ought eagerly to busy himself with continuous use
and enduring exercise in practice, for science by practice
turns into habit; memory and even perception correlate with
the hands and figures, which as an impulse and breath in one
and the same instant, almost the same, go naturally together
for all; and thus will be made a student of habit; following by
degrees he will be able easily to attain this to perfection. And
to reveal more easily the theory I separated this book into xv
chapters,[8] as whoever will wish to read this book can easily
discover. Further, if in this work is found insufficiency or defect, I submit it to your correction.

Leonardo's observation that "this book looks more to theory
than to practice" reads strangely to modern eyes, given the huge
number of practical worked examples the book contains. My
own interpretation of this statement is that he is comparing
Liber abbaci, at least the second edition thereof, with the many
abbacus books that sprang up after his first edition came out,

---

[6]   Leonardo refers here to his book *De Practica Geometriae*, which he published in 1220,
     long after the first edition of *Liber abbaci* had been completed.
[7]   It does not look at all like this to a modern reader.
[8]   Leonardo uses Roman numerals here, since he knows that his readers will not
     understand Hindu-Arabic numerals until they are well past the introduction.

and perhaps also with his own abbacus "book for merchants," the contents of which Franci's 2002 discovery would show to the world. For, without doubt, for all the copious supply of practical examples, *Liber abbaci* is far more than a compendium of numerical recipes. He takes pains to describe the mathematical ideas.

At that point, the prologue changes direction, as Leonardo recounts how he came to learn this remarkable new calculating method. The short passage that follows provides us with the only autobiographical information we have about its author:

> As my father was a public official away from our homeland in the Bugia customs house established for the Pisan merchants who frequently gathered there, he had me in my youth brought to him, looking to find for me a useful and comfortable future; there he wanted me in the study of mathematics and to be taught for some days. There from a marvelous instruction in the art of the nine Indian figures, the introduction and knowledge of the art pleased me so much above all else, and I learnt from them, whoever was learned in it, from nearby Egypt, Syria, Greece, Sicily, and Provence, and their various methods, to which locations of business I traveled considerably afterwards for much study, and I learnt from the assembled disputations. But this, on the whole, the algorithm and even the Pythagorean arcs, I still reckoned almost an error compared to the Indian method.[9] Therefore strictly embracing the Indian method, and attentive to the study of it, from mine own sense adding some, and

---

[9]   Sigler observes here that Pythagorean arcs are mentioned by Gerbert (ca. 980), who became Pope Sylvester II in 999. Gerbert used the Hindu-Arabic numerals on counters, a primitive form of abacus, and marked triples of columns with an arc. These were called Pythagorean arcs. When he writes numbers, Leonardo follows the system of triples, just as we do today when we write numbers like 1,395,281. Leonardo tells his readers that even with various enhancements, abacus methods are no match for Hindu-Arabic arithmetic.

some more still from the subtle geometric art, applying the sum that I was able to perceive to this book, I worked to put it together in xv distinct chapters, showing certain proof for almost everything that I put in, so that further, this method perfected above the rest, this science is instructed to the eager, and to the Italian people above all others, who up to now are found without a minimum.[10] If, by chance, something less or more proper or necessary I omitted, your indulgence for me is entreated, as there is no one who is without fault, and in all things is altogether circumspect.

Why did he include this? After all, like mathematicians before him and afterwards, Leonardo would surely have cared little for the history of the discipline. Mathematics is eternal, and exactly when something new is discovered and by whom is of secondary importance, if indeed it merits observation at all. Mathematicians certainly admire those who make the great discoveries, but their interest is in what is discovered, not in who got there first.

Nevertheless, Leonardo seemed to realize that the invention his book described was a monumental one. At the back of his mind, perhaps not even consciously noticed, may have lurked the notion that one day people would wonder just how this great invention found its way from the Muslim scholars and merchants who had held it for many centuries, into the hands of the practical trading men of northern Europe. In any event, he broke with tradition and inserted an all-too-brief summary of the part he played in the story.

But given what we know about the unreliability of hand-copying of manuscripts, how much faith can we put into what we read?

---

[10] A more colloquial translation of this last clause would be: "Who up to now have lacked this knowledge."

CAN WE BELIEVE WHAT WE READ?
GRIMM CONCLUSIONS

With almost all our knowledge of Leonardo's life coming from one brief paragraph, it would be good to know that what we read is what Leonardo wrote. Unfortunately, we don't have an original manuscript. His words come to us only through copies painstakingly made by hand. And therein lies a problem. Can we be sure that the copies are faithful to the original?

This question was investigated by Richard E. Grimm of the University of California at Davis in the early 1970s. Grimm published his conclusions in a 1973 article "The Autobiography of Leonardo Pisano," published in the *Fibonacci Quarterly*.[11]

In his paper, Grimm examined the brief autobiographical passage in *Liber abbaci* that we just read in English translation. That translation, and most contemporary accounts of Leonardo's life, are based on Boncompagni's 1857 printed Latin transcription of the text, the basis for Sigler's translation. How reliable is the information Boncompagni (and thence Sigler) provided us in that one important passage?

Here is how Grimm summed up the situation:

> His [Boncompagni's] failure to collate his manuscripts and his reliance upon a manuscript often difficult to read led Boncompagni into an astonishing number of errors, both of transcription and of punctuation. The brief autobiographical second paragraph is unfortunately not immune from either type of error; yet this section forms the basis for most of the statements about Leonardo's early life that are found in current histories of mathematics, encyclopedias, and special articles. (p. 89)

[11]   Grimm, 1973.

Grimm began his analysis by creating a new Latin version of the passage based on a collation of the six known manuscripts that contain it. He then provided a heavily annotated translation from the Latin into English. In his notes, he pointed out and explained the ambiguities that arise throughout that one short passage, and the major uncertainties that it left unanswered. What was Leonardo's father's profession and what exactly was he doing in Bugia? How much time did Leonardo spend with him there? What were the ethnicity and cultural attachment of the people who taught him mathematics while there? Exactly what mathematics was he taught there? How long did he spend traveling through North Africa after he left Bugia? Or did he return to Pisa after a brief stay and then make a second trip when he was older? In which case, was it to learn more mathematics or to engage in business?

Finally, Grimm devoted two entire pages to a discussion of the reference in the Boncompagni edition to an "arc of Pythagoras" that Leonardo said he found so lacking. Describing this particular phrase in the translation as "a mare's nest of difficulty which has not been adequately investigated" after a lengthy analysis of the six source manuscripts, Grimm arrived at the somewhat tentative conclusion that a minor error by a scribe corrupted Leonardo's original phrase, which translates as "art of Pythagoras" (*artem* rather than *arcus* in the handwritten Latin text).

Here, according to Grimm, is what Leonardo actually may have written:

> After my father's appointment by his homeland as state of-
> ficial in the customs house of Bugia for the Pisan merchants
> who thronged to it, he took charge; and, in view of its future
> usefulness and convenience, had me in my boyhood come
> to him and there wanted me to devote myself to and be in-
> structed in the study of calculation for some days. There,

following my introduction, as a consequence of marvelous instruction in the art, to the nine digits of the Hindus, the knowledge of the art very much appealed to me before all others, and for it I realized that all its aspects were studied in Egypt, Syria, Greece, Sicily, and Provence, with their varying methods; and at these places thereafter, while on business, I pursued my study in depth and learned the give-and-take of disputation. But all this even, and the algorism, as well as the art of Pythagoras I considered as almost a mistake in respect to the method of the Hindus. Therefore, embracing more stringently that method of the Hindus, and taking stricter pains in its study, while adding certain things from my own understanding and inserting also certain things from the niceties of Euclid's geometric art, I have striven to compose this book in its entirety as understandably as I could, dividing it into fifteen chapters. Almost everything which I have introduced I have displayed with exact proof, in order that those further seeking this knowledge, with its pre-eminent method, might be instructed, and further, in order that the Latin people might not be discovered to be without it, as they have been up to now. If I have perchance omitted anything more or less proper or necessary, I beg indulgence, since there is no one who is blameless and utterly provident in all things.

Apparently, when it comes to learning about Leonardo, we cannot even rely upon what on the face of it appear to be his own words.

CHAPTER 9

Translation

For 800 years, the only way to read *Liber abbaci* (in either handwritten manuscript or typeset book form) was to first learn Latin, the language in which Leonardo originally wrote the text. But all that changed in 2002, when an English language translation appeared. It was, and remains, the only translation into any modern language.

And there we already have a mystery—at least it was so to me. Why did fully—in fact, exactly—800 years elapse between the publication of the first edition in 1202, written in Latin, and the only translation there has ever been into a modern language, the English version, published in the United States in 2002?

When I asked Franci and Pagli about this when I first visited them at the Center for the Study of Medieval Mathematics in Siena, they explained that, not only would translation be an enormous task, but printing would be extremely expensive, since all of the mathematics would have to be laid out in a particular way. "Besides," Pagli added, "there was no need. Only scholars ever need to read it, and they can all read Latin."

Fortunately for me, Laurence Sigler had come to a different conclusion, and had labored to produce an English translation.

After I read Sigler's book, I intended to contact him to ask about the translation process—what motivated him to take on such a mammoth task, how long had it taken, what were the particular challenges, and above all, what impressions did he gain of Leonardo the man.

After all, Sigler had lived with Leonardo's words (in the original Latin) for so long, he surely had a better sense of the mind of Leonardo than anyone else alive today. You can imagine then my shock, and sadness, when I opened the book, freshly delivered from Amazon.com, and saw the translator listed as "Laurence Sigler (deceased)."

A brief introductory note on the first page was signed "J. M. Sigler." Sigler himself had died before the book was published—and with him went any insights he might have developed into the nature of the book's author. But maybe this "J. M. Sigler" could help me. I was about to embark on yet another twist in the remarkable story of *Liber abbaci*.

Sigler had been a professor in the Mathematics Department at Bucknell University in Lewisburg, Pennsylvania, and the book's introductory note by J. M. Sigler credited several of his former colleagues there with helping to arrange publication after his death. I contacted one of them, Greg Adams, by email. In his reply, in addition to giving me a brief outline of the (remarkable) events surrounding the publication of the translation, Adams put me in touch with the author of the introductory note, Laurence Sigler's widow, now remarried as Judith M. Sigler Fell. In a brief email exchange, Judith Sigler told me a little about her late husband.

A first-rate mathematician, Laurence ("Larry") Sigler studied Physics and Mathematics at Oklahoma State University, went on to obtain a PhD in mathematics at Columbia University, and then

FIGURE 11. Laurence Sigler, *Liber abbaci* translator, in his office at Bucknell University. Photograph courtesy of Judith Sigler Fell.

spent a year in Cambridge, England as a Fulbright scholar. He joined the faculty at Bucknell University in 1965, and remained there for the remainder of his career. He wrote several textbooks in algebra and set theory before turning to the daunting task of translating the works of Leonardo.

He completed the first of these translations, *Liber Quadratorum*, in 1991, after five years work. By the time that translation was published the following year, he was already hard at work on the second, far more ambitious, translation, that of *Liber abaci*.[1] It was to be his last one. While he was in the middle of the project, he learned that he was terminally ill with cancer, and he spent the final years of his life in a desperate fight against time to finish the job. He died in 1997. He managed to complete the translation, but did not have time to finalize the publication details. Judith Sigler

---

[1] For some reason, Sigler used the more common spelling of *abaci*, with one *b*.

decided to handle that final task. It would turn out to be a difficult five-year struggle.

What she and Greg Adams told me about the publication process was so amazing—in fact Adams used the very adjective "amazing" in his brief email description—that I knew I had to travel to Lewisburg, the small Pennsylvania college town where Bucknell University is located, to talk with both Adams and Sigler's widow. My plan was not only to learn more about Laurence Sigler and his work on the translation, but also to hear from Judith what was involved in getting his translation finished and into print. For Judith Sigler too was a major player in the historical saga of *Liber abbaci*.

What I would learn from that visit was that the story of *Liber abbaci* is a very human one, spanning many centuries, with an ending (assuming the translation into English is its ending) every bit as dramatic as any Hollywood scriptwriter could dream up.

Unfortunately, my university duties meant I had to put that part of *my* Leonardo project to one side, and with one thing leading to another, it was not until 2009 that I was finally able to make the trip east and learn the remarkable full story of how *Liber abbaci* came to be translated into English.

As it turned out, putting my project on hold for several years did have one positive benefit for me. It gave me time to reflect on the degree to which the story of how numbers and arithmetic reached—and impacted—the West depended so often on the simple acts of copying and translating manuscripts by individual human beings.

Transmission of our number system and its arithmetic from the Hindus, who originally developed it, to the Arabs, who extended it and made practical use of it, depended on translations of manuscripts from Hindi to Arabic. To become accessible to European scholars, a manuscript first had to be translated again into Latin. Hindu-Arabic arithmetic achieved widespread use by Italian traders only after manuscripts were translated into vernacular Italian. For the system to gain worldwide adoption, those manuscripts then had to be further translated into many different languages.

Sigler's first-ever translation of *Liber abbaci* into a modern language, published in 2002, exactly 800 years after Leonardo published his first version of the book, may not change the course of history the way those earlier, pivotal translations did. But it is part of the entire story of those translations. Its significance is a historical one, allowing us to understand the journey of some of the key ideas that made us what we are today.

What is uncanny is how, even in the present-day world of photocopy machines and computer typesetting of manuscripts by authors, the very same problems that bedeviled the transcribers and translators of much earlier times can still arise.

Judith Sigler

During the nighttime hours of Monday December 7, 2009, a major snowstorm in the Midwest meant that my Tuesday morning flight from San Francisco to Chicago and then on to Harrisburg, Pennsylvania was canceled. But as the morning drew on, the storm abated and I was eventually able to secure a seat on two later flights, arriving in Harrisburg late in the evening.

I had reserved a room at a large, multistory, business hotel close to the airport. I must have checked in to similar hotels hundreds of times during my career. They all look much the same, they all provide the same level of comfort, they all offer the same amenities, and they all serve the same kind of food. Exactly what you are looking for if you travel a lot on business, as I do.

The Best Western Harrisburg Inn proved as easy to find as their website claimed, less than five miles from the airport. It stood in the middle of its own, large parking lot, around the edge of which were scattered a gas station and several major-chain, fast-food restaurants. My trip out having been delayed by many hours, I was tired and hungry as I drove my car onto the lot. A quick meal, probably in the hotel since the temperature was close to freezing, and then straight to bed.I found a parking spot fairly close to the front entrance, put on my coat, took my bag off the back seat, and started to walk toward the hotel. At that very moment (I'm not making this up, not even embellishing it for effect), the night air filled with loud sirens, and from every conceivable direction, fire trucks and police cruisers raced into the parking lot, their flashing red and blue lights illuminating the entire area. Hotel guests began streaming out of every exit. The first team of firefighters rushed into the lobby exactly as I reached the door. I looked in and could see the check-in desk, with the clerk standing behind it, speaking on the telephone. I could tell that one of the sirens was the hotel fire alarm—I hadn't been able to hear it from where I had parked.

With a large, fully outfitted firefighter standing between me and the warmth of the hotel, it was clear I was not going to set foot inside the building, and a few moments later all the guests were ushered away from the semi-warmth of the wedged-open front door and out into the parking lot.

With more fire trucks and police continuing to arrive—there must have been a dozen by now and I could hear more sirens approaching—it was clear that it would be some time before I could get to my room. If at all. The historical research business is not always a comfortable activity, I realized.

After putting my bag back in the car, I traipsed across the parking lot, stepping over fire hoses and walking cautiously between all the emergency vehicles, to the Taco Bell, which was the closest source of warmth and food, to wait until the emergency was over.

Roughly an hour and a couple of Americanized tacos later, the fire trucks started to leave, and I was able to gain access to my room. Apparently a videogame machine in the recreation room had caught fire.

As incidents go, it turned out to be pretty minor, and for me just a slight inconvenience. I would not have bothered to mention it were it not for the fact that events have seemed to conspire against me throughout my entire Leonardo quest. If I were superstitious or believed in the supernatural, I might have started to feel nervous. But I'm not, so I didn't. I was just pleased to be one step closer to finishing my quest and publishing my book about Leonardo.

After spending the night at the hotel, early the next day I drove myself to Lewisburg, the melting snow slowing the progress of my small rental car, with the result that the 70- mile journey took almost two hours.

Thanks to Google maps and directions, I had no trouble finding the fairly spacious, attractive house where Judith Sigler lives, on a quiet lane looking across State Highway 15 to the campus of Bucknell University. She and her husband, Joe Fell, a retired philosophy professor from Bucknell, both came to the door to greet me as I drove up.

Judith Fell is a somewhat short woman with a cheery face and dark hair, whom I took (correctly) to be about the same age as me. She was wearing grey jeans and a polo-necked sweater with horizontal stripes of soft blue, grey, green, and yellow. After we spent a few moments getting to know each other, I asked her to tell me about herself. Eager to focus on her late husband, she said she preferred we talk about him, but later on, at my insistence, she did tell me about her life.[2]

---

[2] As you will shortly discover, this was not idle curiosity on my part. As I hinted earler, Judith Sigler is not just the narrator of the story you are about to read. She is an integral part of it.

She was born Judith von Szäger, in Budapest, Hungary, in 1946. An unplanned child to older parents, she was conceived shortly after her father, an engineer, returned home after his release from a Russian prisoner-of-war camp at the end of the Second World War.

Judith grew up during a golden age of Hungarian research in physics and materials science. One of the dominant, state-owned engineering companies was Tungsram, where her father worked. Although she herself later became a qualified engineer, her first choice had been to study biology. Her desire to pursue research into living organisms was motivated by the death of her elder brother. Hungary had just purchased a cobalt gun, and three months after being exposed to its invisible, deadly rays, Judith's brother became one of the world's first victims of commercial radiation poisoning.

Prevented from majoring in biology by the strategically motivated quota system the communist Hungarian government imposed on university courses, she had to elect engineering. As a result, she eventually found herself working on the newly emerging technology of computers.

As a member of the Hungarian intelligentsia, she got to know a number of mathematicians, among them the famous Paul Erdös, who I also knew fairly well.

In 1975, she left Hungary, living for three years in Monaco before moving to the United States in 1978. As she did not speak English, Judith first supported herself by taking a job as a maid, later advancing to be a nanny, where, she told me, she benefited from the best language teacher she ever had: her 4-year-old charge. After that, with her expertise in transistor technology outdated, she was unable to secure a position as an engineer, and she worked as a sales representative for a jewelry casting company in New York.

In 1979, she met Laurence Sigler, the two of them married, and she moved to Lewisburg. There, she studied accounting at

Bucknell, and began to work as a tax accountant, a profession she still engaged in when we met.

LAURENCE SIGLER

Laurence Sigler—all his Bucknell colleagues called him Larry, though to Judith he was always "Lorenzo"—was born in Tulsa, Oklahoma in 1928. He studied mathematics and physics at Oklahoma State University in Stillwater, and went on to complete a PhD at Columbia University, New York, under the guidance of Walter C. Strodt. His dissertation was titled *On the Real Asymptotic Theory of the Factorization of Ordinary Differential Operators*, for which he was awarded his PhD degree in 1963. After graduating from Columbia, he spent a year at Jesus College in Cambridge University in England, on a Fulbright Scholarship. After that, he held short-term positions at Hunter College and Hofstra University, both in New York, before going to Bucknell University in 1965. He remained there, in rural Pennsylvania, until his death in 1997. He wrote a number of textbooks, among them an algebra textbook, published in 1978, and *Exercises in Set Theory*, published a year earlier as a companion to the popular set theory textbook by the famous emigre Hungarian mathematician Paul Halmos. (The Hungarian edition was published as a single volume under their joint names.)

Sigler became interested in Leonardo by chance, Judith told me. He set out to write a college-level mathematics book that would interest non-science majors, and the Fibonacci sequence is an obvious topic to include in such a book. When he looked for information about Fibonacci, he discovered that almost nothing was known. Moreover, there were no English translations of any of Leonardo's books. In fact, the only modern language translation of anything Leonardo had written was a French edition of the *Liber Quadratorum*.

Frustrated at this state of affairs, Sigler decided to do something about it. The college textbook project was put aside, and he set about teaching himself Latin in order to translate—with annotation—Leonardo's three surviving major works into English, beginning with *Liber Quadratorum*, then *Liber Abbaci*, and finally *De Practica Geometrie*.

And so, in the early 1980s, armed only with a small paperback edition of *Cassell's Compact Latin Dictionary*, this self-taught Latin scholar and mathematical historian began work on the first translation. He completed it in 1986 for publication the following year by Harcourt Brace. With that first translation completed, he started at once on the far more daunting task of translating *Liber abbaci*, the second Leonardo book on his list.

It was to be his last one. In 1992, when the project was nearing completion, he learned that he was terminally ill with lymphocytic leukemia, and he spent the final years of his life in a desperate fight against time to finish the job and get the book published. Because the disease results in a compromised immune system, he was advised to minimize his exposure to virus infections, and so he retired from the university to focus entirely on the book. Although he was 64, an age at which many of his healthy colleagues were happy to retire, Sigler, who had taught a full course load throughout his career at Bucknell, regarded this as "early retirement."

With a heightened sense of urgency due to his deteriorating health, he finally completed the *Liber abaci* [3] translation in 1995, copied the files onto a dozen or so magnetic floppy disks, and sent them to the editor who had published his earlier translation.

Unfortunately, at that time, the editor was in the process of leaving Harcourt Brace to establish his own scientific publishing enterprise, and wrote back to say he wanted to delay publication

---

[3]   Remember, Sigler used the spelling with a single "b."

some months and then handle the project through his new company. With the book still not under formal contract and Harcourt now out of the picture, the manuscript was therefore in the hands of just one man, who was in the midst of creating his new company.

As the two corresponded over the following months, Sigler grew more anxious, wanting to see the translation published before he died. Knowing nothing of this particular urgency, however, the publisher continued to wait until he had the resources to handle a project as complex as *Liber abaci*. The book was still sitting at the publisher's, awaiting publication, when Sigler died in 1997.

PUBLICATION

When Laurence Sigler died, Judith decided she would complete the work for him, by handling the final publication details, both to fulfill his wishes and as a testament to his memory. But there was a twist. Worried that if the publisher learned that Laurence had died, he would no longer publish the book, she made the unusual decision to act not on behalf of her just deceased husband, but *as* him.

She took over his university email account and started to communicate with the editor as "Larry Sigler." (To this day, she still uses Laurence Sigler's original email account at the university. Such is the warmth the members of the Bucknell Mathematics Department feel toward Larry and Judith, that they are more than happy for Larry's presence to continue in this way.)

Months passed, then years, but still the publisher offered no date for publication. With her subterfuge restricting her to communication by email, Judith did not know what to do to ensure the book's appearance.

One day in 2000, she mentioned the difficulty she was having to an old friend of Larry and hers who was visiting, Alex Khoury,

FIGURE 12. Laurence and Judith Sigler. Judith prepared her late husband's translation for publication after his death. Photograph courtesy of Judith Sigler Fell.

a mathematics professor from the University of Michigan. After looking at the manuscript, Khoury told Judith he would do all he could to find another publisher, this time a well-established one with a good reputation, as quickly as possible.

Judith was not convinced this was a wise move, but then Alex secured the assistance of the Bucknell mathematicians, who managed to persuade her to switch publishers. In particular, Bucknell mathematician George Exner knew an editor at the prestigious, and much bigger, Springer-Verlag, a leading publisher of mathematics books, and he approached Springer on Judith's behalf. Springer replied immediately, saying they would publish the book speedily.

In his original email to me back in January 2003, Greg Adams had described the sequence of events that followed as "amazing." From my lengthy conversation with Judith Fell, I discovered he had not been exaggerating.

Springer had just two initial requirements before they could proceed with publication. The first, and most pressing, was they needed to have the manuscript in electronic file format.

Unfortunately, the only electronic version was stored on the floppy disks held by the first publisher, the one who had left Harcourt Brace. So Judith wrote to him, telling him that Laurence had died, and that she wanted the electronic files back in order to submit the book elsewhere. He complied at once, and returned the disks.

But three years had elapsed since Laurence had sent off the manuscript, and the technology had changed. Neither Judith nor anyone at Bucknell had a computer that was able to read the files on Laurence's disks. At that point, Judith asked Joe Fell's daughter Caroline if she could help; at the time she worked for Pearson Publishing in New York. Caroline suggested they send the disks to a computer-savvy friend from her student days at Yale, now living in Cambridge, who might be able to hack the disks and retrieve the information on them.

The hacker (Judith would not tell me his name) was unable to recover the files in their entirety, but he was able to retrieve the unformatted text. Lost forever, however, were all of the mathematical layout instructions and the graphics. Judith would have to reconstruct them based on the paper, working copy of the manuscript Larry had left behind. That consisted of a thick pile of pages having a three-lined structure, with the Latin text on the top line, which Laurence had typed in by hand from the photocopy of the Boncompagni manuscript he had been working from, his handwritten translation notes on the line beneath, and his final translation handwritten below that.

Even worse, when Judith looked at the recovered text more closely, she discovered that around 80 pages were missing. Presumably one of the disks had been lost. She would have to recreate that part of the file afresh from Laurence's three-lined translation manuscript. Still, although she was not a mathematician, she had been an engineer, and the mathematical content of *Liber abbaci* was, after all, basic arithmetic and algebra, so she felt sure she could handle it.

Then there was Springer's second requirement. Not only did they need the work as an electronic file, it had to be formatted using a particular system: the mathematical typesetting language TeX (see momentarily).

In setting out to translate *Liber abbaci*, Sigler had faced the same problem that any author of a mathematics book has to deal with: How do you handle the mathematical symbols? The problem is particularly acute if the text is dotted with complicated formulas. Mathematical notation evolved as a system for writing mathematics by hand. Whereas ordinary text lies on a horizontal line[4] that you read in a simple left-to-right fashion,[5] in mathematical formulas the overall layout on the page can be important, vertical placement as well as horizontal. The relative sizes of the different symbols can also be significant, and so can the typeface in which a symbol is set—normal type, italics, boldface, or a fancy, Gothic-like style. Moreover, mathematicians use Greek and Hebrew letters and a variety of other special symbols. Until the advent of modern word processing systems, authors of mathematics textbooks used to type (or have typed) the basic text of their manuscripts, leaving spaces where they would later insert the symbols and formulas by hand. The publisher would then have to send the manuscript to a specialist printing shop, to be typeset by experts in such work.

Liber abbaci does not have any special symbols; it is a book about basic arithmetic and elementary (non-symbolic) algebra. But it does have a great many examples showing how to do arithmetic

---

[4]   Or a vertical line for some languages, such as Chinese.
[5]   Or right-to-left for Arabic, Hebrew, and some other languages.

and solve problems using algebraic methods. Leonardo's main purpose in writing *Liber abbaci* was to show people how to write numbers in Hindu-Arabic notation, how to lay them out in order to do arithmetic. As everyone today knows, doing arithmetic effectively involves arranging numbers in rectangular arrays, with digits correctly lined up vertically as well as horizontally. Sigler had to find a way to incorporate such arrays into his manuscript.

During the 1970s, as word processing computer programs started to become more widely available, two different methods were developed to enable authors of mathematics books and papers to prepare their manuscripts for publication. In one method, initially the most popular, software manufacturers took ordinary word processing packages and added a variety of special fonts that allowed the user to assemble a mathematical formula from basic building blocks, presented in a select-and-paste tabular menu. Such systems were easy to use, but it was slow, painstaking work to select the various elements and use the mouse and cursor to put them in the right positions. Moreover, the results were rarely aesthetically pleasing, with irregular spacings, superscripts and subscripts that were too large or too small, and so forth. But at least these systems enabled authors to compose their own manuscripts with no special training.

The other approach was developed by the Stanford-based computer scientist Donald Knuth. His system, called *TeX* (pronounced *tek*; the unusual capitalization is part of the brand name), produces correctly laid-out mathematical text, both words and formulas, with even the most complex of mathematical formulas perfectly typeset. Knuth based his system on the simple idea of regarding a page of written mathematics not as text but as a picture.

Authors who use TeX type the words of the manuscript in the normal way, but when they want to insert a formula, they type instructions that tell the system how to draw the formula as a picture. This means that the user has to learn—or keep looking up—all the commands that must be typed to create the various symbols. This makes TeX an extremely difficult system to use at first, but once a user has mastered it sufficiently well, preparing mathematical manuscripts becomes fairly routine, no matter how complicated the formulas are. The TeX program has in its memory all the rules for composing mathematical formulas and laying them out properly. The author uses an ordinary word processor to prepare a regular text file, which contains no mathematical symbols (other than numbers).

The author does, however, have to type commands for creating mathematical symbols and formulas. [TeX commands are indicated by a backslash prefix. For example, \int tells TeX to draw an integral sign, \infty tells it to draw an infinity symbol, and \frac{\pi}{4} tells TeX to draw the fraction $\pi/4$ in proper mathematical fashion with the π placed above the 4 and a horizontal line between them.]

Sigler originally typed his translation using a program called *MathText*, a system of the former kind. This was the encoding that the un-named hacker in Cambridge had to break in order to recover the raw text.

Judith now had to take the recovered text and turn it into a fully functional TeX file. This was an enormous challenge. In effect, she would have to become yet another translator of *Liber abbaci*, this time translating her late husband's English translation of Boncompagni's nineteenth-century printed version of a thirteenth-century handwritten copy of Leonardo's original handwritten Latin manuscript into the computer language TeX.

Judith told me that were it not for the support and help she was offered—and given—by Larry's former colleagues in the Bucknell Mathematics Department, she would not have undertaken this second task. TeX is an extremely powerful, highly sophisticated technical typesetting language that requires considerable time and

FIGURE 13. The Bucknell University mathematicians who assisted Judith Sigler in publishing her late husband Laurence's English translation of *Liber abbaci* after his death. *Left to Right*: Paul McGuire, Department Chair when work was completed; Greg Adams, initiated the help, on sabbatical during completion; Judith Sigler Fell, née von Szäger; George Exner, who knew Thomas Voerster at Springer; Karl Voss, Department Chair when I visited, new at the time, whose TeX skills were crucial.

effort to master. And typesetting *Liber abbaci* would clearly require considerable mastery. (I have been using TeX to write articles and books for almost 30 years now, and I do not know how to achieve some of the special layouts required for *Liber abbaci*, that Judith eventually produced.)

Greg Adams, who had promised her his support, was away on sabbatical when Judith began teaching herself TeX and working on Laurence's manuscript. But the department chair, Paul McGuire, was equally supportive, allowing Judith the use of his office and computer (which had the TeX system on it) at nights, when all the faculty and students had gone home. By good fortune, the department had just hired a young mathematics professor named Karl

Voss (he was the new department chair when I visited in 2009), and, like most younger mathematicians, he was a TeX expert, and thus able to help Judith with some of the more difficult typesetting issues. He also, Judith told me, came in late at night to rescue her when she phoned him to say Paul's computer had crashed.

According to Greg Adams, Judith not only rapidly became expert in using TeX, but also in knowing the contents of *Liber abbaci*. In particular, in consultation with Springer, the publisher, she made typesetting changes that greatly improved the readability of the text, such as moving certain things to the margins. Using her husband's notes, she was also able to overcome the Springer editor's reluctance to include some of Laurence's introductory remarks.

And then, finally, the task was completed. It had taken six months of intense effort, from the beginning of April to the end of September 2001. Judith sent the completed manuscript to Kristina, Larry's daughter (by his first marriage), who had graduated from Penn State with degrees in History and English, to proofread, before mailing it to Springer. As Judith left Paul McGuire's office with the package containing the disks, she was met by the entire department, who had lined up on both sides of the corridor. They had come to applaud her achievement and accompany her to the campus post office, so they too could witness Larry's translation finally head out to the world. By pure chance, publication would be in 2002, exactly 800 years after Leonardo completed his first edition of *Liber abbaci*.

The world finally had a version of *Liber abbaci*, one of the most influential books ever written, in an international modern language. But it had taken an amazing sequence of events, and the efforts of a number of people, to produce it. In particular, as Adams had remarked of Judith Sigler in his 2003 email to me, "While she had help from various people in the [Bucknell mathematics] department, her achievement was nothing short of remarkable."

POSTSCRIPT

In January 2003, shortly after Sigler's English translation was published, and five years after his death, stuffed rabbits began appearing in a display case in the Bucknell mathematics department—one on the first day, a second the following day, then the next day there were 3, followed by 5, 8, 13, and 21. This is the famous Fibonnaci sequence, in its original formulation, as the size of a growing rabbit colony. Laurence Sigler's colleagues had found a unique way to celebrate the mammoth labor of love—both by Laurence in the final years of his life, and by his wife Judith after he had passed away—in giving the world its first, and only, modern-language translation of *Liber abbaci*. People who think that mathematics is dry and boring, and mathematicians even more so, clearly have not met any real mathematicians.

FIGURE 14. The Display Case in the Bucknell Mathematics Department, celebrating the publication of Laurence Sigler's translation of *Liber abbaci*, 800 years after the first edition appeared. Photograph courtesy of Judith Sigler Fell.

CHAPTER 10

Reading Fibonacci

Thanks to Larry Sigler and his widow Judith Fell, and the mathematics faculty at Bucknell University who helped her, a few months after I returned to Stanford from my trip to Italy, I was able to open up an English-language edition of *Liber abbaci* and start to read it.

While millions of people around the world were poring over the new Harry Potter book—which had advance orders from Amazon numbering in the tens of thousands—I suspected I was the only person on the planet (likely correct apart from Professor Goetzmann of Yale) who was starting to work through *Liber abbaci*. But I guarantee you that there is no way my excitement could have been any less than that of J. K. Rowling's myriad readers.

The first seven chapters of *Liber abbaci* focus on the basic mechanics of Hindu-Arabic arithmetic (starting with how to write and manipulate the numbers themselves, in Chapter 1), which Leonardo explains using (many) specific numerical examples, much like the way elementary school pupils are taught today. Our familiar educational practice (for older students) of summarizing the various rules by simple algebraic formulas came centuries later.

It is with Chapters 8 and 9 that the reader first encounters real-world examples. Many of these involve items called "rolls," a roll being a unit of weight, equal to 12 ounces. However, units of weight differed from one city to another. So you had "Pisan rolls," "Florentine rolls," and so forth. One worked problem in Chapter 8 is titled:

On finding the worth of Florentine rolls when the worth of those of Genoa is known. [p. 148]

A typical worked problem in this section of the book starts like this:

If one hundredweight of linen or some other merchandise is sold near Syria or Alexandria for 4 Saracen bezants, and you will wish to know how much 37 rolls are worth, then . . . [p. 142]

The chapter on companies and their members (Chapter 10) demonstrates obviously valuable methods for solving problems such as determining the payouts in the following scenario:

Three men made a company in which the first man put 17 pounds, the second 29 pounds, the third 42 pounds, and the profit was 100 pounds. [p. 220]

Toward the end of Chapter 11, we find a curious problem that became quite well known to mathematicians (though not as famous as his "rabbit problem," which I'll get to shortly). It is called "Fibonacci's Problem of the Birds." Here is what Leonardo asks:

ON A MAN WHO BUYS THIRTY BIRDS OF THREE KINDS
FOR 30 DENARI

A certain man buys 30 birds which are partridges, pigeons, and sparrows, for 30 denari. A partridge he buys for 3 denari, a pigeon for 2 denari, and 2 sparrows for 1 denaro, namely 1 sparrow for ½ denaro. It is sought how many birds he buys of each kind. [p. 256]

What makes this problem particularly intriguing is that, on the face of it, you don't have enough information to solve it. You arrive at that conclusion as soon as you try to solve it using modern symbolic algebra. If you let *x* be the number of partridges, *y* the number of pigeons, and *z* the number of sparrows, then the information you are given leads to two equations:

$x + y + z = 30$ (the number of birds bought equals 30)

$3x + 2y + \frac{1}{2}z = 30$ (the total price paid equals 30)

But as everyone learns in high school algebra class, you need three equations to find three unknowns. Well, in general that is true, but in this case you have one crucial additional piece of information that enables you to solve the problem. Do you see what it is? I'll give the solution to the problem at the end of this chapter. (Leonardo, as usual, presents the solution in words, not symbols, but apart from that, the solution I will give you is his.)

Not all of Leonardo's problems are presented in obviously practical terms. For instance, many are like this one in Chapter 12 of *Liber abbaci*:

A certain lion is in a certain pit, the depth of which is 50 palms, and he ascends daily 1/7 of a palm, and descends 1/9. It is sought in how many days will he leave the pit. [p. 273]

But for the most part, the examples of *Liber abbaci* are couched in self-evidently practical terms. In fact, Chapter 12 is a mammoth piece of work that presents 259 worked examples, each of which Leonardo works through in great—I would say gory—detail. Some examples require only a few lines to solve, while others spread over several densely packed pages. In Sigler's English-language translation, the entire chapter takes up 187 printed pages.

When I first read *Liber abbaci*—no, looked through would be more accurate, its sheer volume made it too daunting a task to really *read* it—I could not believe that a really smart guy like Leonardo, arguably the brightest mathematician of his day, would devote his time to what must have been an unbelievably tedious task of spelling out in such minute detail one calculation after another. After all, each one differed only slightly from many others. But then it struck me.

The appearance of *Liber abbaci* was very similar to the birth of the personal computer in the early 1980s in the revolutionary effects both had on the societies of their days.[1] The young computer whizzes who developed the personal computers and the software that ran on them spent hours poring over the fine details of circuit designs and software code. By any measure but one, nothing could be more tedious and mind-numbing. The one exception will be familiar to anyone who has stayed up all night writing a computer program. Once you get into the project, it develops a life of its own. You find yourself in what is often referred to as "the flow." Time stands still, and the mind is able to cope with any amount of fine detail. Indeed, it does not seem like fine detail; at that moment that design or that piece of code is all that matters in the world.

This is particularly so if you believe that what you are doing is important—so important it could change the world. To the reader today, like me, Leonardo's text describes something I have been familiar with since my early childhood math classes. But at the turn of the thirteenth century, none of it was known. When Leonardo was writing his mammoth work, it was as new to him as to his future readers. In many cases he was working out the examples

[1]   I will elaborate on this observation in chapter 14.

for the first time ever. He was working them out for himself as much as for his later readers. The methods he was developing and using were as new and fresh to him as the silicon-chip specifications and computer programs were to Bill Gates (who acquired and developed the MS-DOS operating system that ran the first IBM PC), Steve Wozniak (who designed the Apple I and Apple II computers), Dan Bricklin (who wrote VisiCalc, the first computer spreadsheet), and all the other pioneers of the personal computer revolution. And, like them, Leonardo surely sensed that what he was doing was tremendously important—something that would change life on Earth forever.

Adding Spice

Chapter 12 is particularly long, a huge compendium of worked problems. Like mathematics teachers and authors before him and since, Leonardo clearly knew that many of the people who sought to learn from him would have little interest in theoretical, abstract problems. Though mathematicians feel entirely at home in a mental world of symbols, most people prefer the more concrete and familiar. And so, in order to explain how to use the new methods he learned during his visit to North Africa, Leonardo looked for ways to dress up the abstract ideas in familiar, everyday clothing. The result is a class of problems that today go under the name "recreational mathematics."

For instance, he presents a series of "purse problems" to try to put into everyday terms the mathematical problem associated with dividing up an amount of money—or anything else that people may want to divide up—according to certain rules. The first one goes like this:

Two men who had denari found a purse with denari in it;
thus found, the first man said to the second, If I take these

denari of the purse, then with the denari I have I shall have
three times as many as you have. Alternately the other man
responded, And if I shall have the denari of the purse with
my denari, then I shall have four times as many denari as you
have. It is sought how many denari each has, and how many
denari they found in the purse. [p. 317]

Students today would be expected to solve this problem using
elementary algebra (equations), and it takes at most a few lines.
But modern symbolic algebra is a much later invention. Leonardo
filled almost half a parchment page with his solution. (At heart,
it is the same solution as today's algebra student will—or at least
should—come up with, but without the simplicity of symbolic
equations it takes a lot more effort, and a great deal more space
on the page, to work through to the answer.)

More complicated variations follow, including a purse
found by three men, a purse found by four men, and finally a
purse found by five men. Each problem took a full parchment
sheet to solve. Adding still more complexity, he presented
a particularly challenging problem in which four men with
denari find four purses of denari, the solution of which fills
four entire pages of Sigler's English-language translation. In
all, Leonardo presented 18 different purse problems, which
occupy nineteen-and-a-half pages of the English-language
translation.

Although many of the variants to the purse problem that
Leonardo presents seem to be of his own devising, the original
problem predated him by at least 400 years. In his book *Ganita Sara
Sangraha*, the ninth-century Indian mathematician Mahavira (ca.
800–870) presented his readers with this problem:

Three merchants find a purse lying in the road. The first as-
serts that the discovery would make him twice as wealthy as
the other two combined. The second claims his wealth would

triple if he kept the purse, and the third claims his wealth
would increase fivefold.

The reader has to determine how much each merchant has and
how much is in the purse. This is precisely Leonardo's first purse
problem in Chapter 12 of *Liber abbaci*. Presumably Leonardo came
across the puzzle by way of an Arab text.

Leonardo's purse problems involve divisions that require only
whole numbers. To explain how to proceed when fractions are
involved, he used a different scenario his readers could relate to:
buying horses. On page 337 of Sigler's translation, we read:

> Here Begins the Fifth Part on the Purchase of Horses among
> Partners According to Some Given Proportion.

The first horse problem reads:

> Two men having bezants found a horse for sale; as they
> wished to buy him, the first said to the second, If you will
> give me 1/3 of your bezants, then I shall have the price of
> the horse. And the other man proposed to have similarly the
> price of the horse if he takes 1/4 of the first's bezants. The
> price of the horse and the bezants of each man are sought.
> [p. 337]

Again, a math student today would solve this problem using (symbolic) algebraic equations. Leonardo solved the problem using
arithmetic. As a result, this is where I find myself glossing over
the page. I suspect you will too, but for the record, here is what
Leonardo actually wrote.

> You put 1/4 1/3 in order, and you subtract the 1 which is
> over the 3 from the 3 itself; there remains 2 that you mul-
> tiply by the 4; there will be 8 bezants, and the first has this

many. Also the 1 which is over the 4 is subtracted from the 4; there remains 3 that you multiply by the 3; there remains 9 bezants, and the other man has this many. Again you multiply the 3 by the 4; there will be 12 from which you take the 1 that comes out of the multiplication of the 1 which is over the 3 by the 1 which is over the 4; there remain 11 bezants for the price of the horse; this method proceeds from the rule of proportion, namely from the finding of the proportion of the bezants of one man to the bezants of the other; the proportion is found thus. [p. 337]

This does not read like a modern-day arithmetic textbook. It is far more reminiscent of a cookery recipe, perhaps one found in a book titled *Basic Cookery for Dummies*, written to leave nothing to chance. Leonardo explains, step-by-step, what digits you must write where, and what you do to them. A modern textbook would supply an algebraic formula into which you can simply plug numbers, but algebraic notation was still several centuries away. Instead, Leonardo had to convey the method by giving many concrete examples, each with its unique twist and each using slightly different numbers.

Thirty-six pages and twenty-nine horse-type problems later, Leonardo evidently decided he had provided enough variations that his readers will have mastered the general technique. Along the way, he worked out a problem in which five men buy five horses [p. 350], another, particularly tricky puzzle that he titles

A Problem Proposed to Us by a Most Earned Master of a Constantinople Mosque [p. 362]

in which five men buy not a horse but a ship, and another problem where seven men buy a horse, which, despite its seemingly greater complexity, turns out to be less intricate to solve [p. 366].

Since many of his fellow citizens were frequent travelers, Leonardo knew that money problems about traveling were sure to attract wide interest, so these make up his next set of examples. For his first traveler problem, he wrote:

> A certain man proceeding to Lucca on business to make a profit doubled his money, and he spent there 12 denari. He then left and went through Florence; he there doubled his money, and he spent 12 denari. Then he returned to Pisa, doubled his money, and spent 12 denari, and it is proposed that he had nothing left. It is sought how much he had at the beginning. [p. 372]

While the ending of his little scenario might strike a familiar chord to many a vacation traveler to Tuscany today, this particular problem has a relatively easy solution. So too do some, though not all, of the many variants of the problem that Leonardo solves in the ensuing pages. He also illustrates the same arithmetical principles and solution methods with some other problems, including several about calculating interest on house purchases [pp. 384–92].

One problem leads to the particularly nasty answer that a certain businessman walks away from a partnership in Constantinople with a profit of

$$\frac{1\ 7\ 1\ 4\ 0\ 21169}{2\ 8\ 8\ 8\ 8\ 24767}\ 206 \text{ bezants}$$

To read this, you need to understand that when Europeans in Leonardo's time learned the Hindu-Arabic number system they wrote fractions before the whole number part, building the fraction up from right to left, with each new fraction representing that part of what is to the right. For example,

$$\frac{1\ 2\ 4}{2\ 3\ 5} \text{ means} \frac{1}{2\times3\times5} + \frac{2}{3\times5} + \frac{4}{5}, \text{ i.e. } \frac{29}{30}$$

The right-to-left ordering may simply be a carryover from the writing of Arabic, although it is of interest to note that, for the most part, Arabic texts expressed Hindu-Arabic numbers rhetorically, using words instead of symbols. Leonardo would have articulated the above fraction as the Arabic mathematicians would both write and speak it: "Four fifths, and two thirds of a fifth, and one half of a third of a fifth."

Decimal expansions are a special case of this notation when the denominators are all 10. For example, Leonardo would have written today's decimal number 3.14159 as

$$\frac{9}{10}\frac{5}{10}\frac{1}{10}\frac{4}{10}\frac{1}{10}3$$

Though decimal representation seems far simpler to us today, there was little need for it in Leonardo's time, as no one counted anything special in tenths. In fact, the method used to represent fractions was particularly well suited for calculations involving money. The monetary system in use in medieval Pisa was essentially the same as that used in the UK until 1970, with 12 denari equal to 1 soldus and 20 soldi equal to 1 libra. Thus 2 librae, 7 soldi, and 3 denari would be written

$$\frac{3}{12}\frac{7}{20}2$$

Units of weight and measure could be even more complex. According to Leonardo, Pisan hundredweights[2]

> have in themselves one hundred parts each of which is called a roll, and each roll contains 12 ounces, and each of which weighs ½ 39 pennyweights; and each pennyweight contains 6 carobs and a carob is 4 grains of corn.

Imagine having to calculate with those units.

2   Sigler, 2002, p. 128.

Interestingly, an Arabic arithmetic text written by al-Uqlidisi in Damascus in 952 did in fact use place-value decimals to the right of a decimal point, but no one saw any particular reason to adopt it, and so the idea died, not to reappear again for 500 years, when Arabic-speaking scholars picked up the idea once more. Decimal fractions were not used in Europe until the sixteenth century.

Fractions written after the whole number part in Leonardo's time denote multiplication. For example, ½ of 3.14159 could be written

$$\frac{9}{10}\frac{5}{10}\frac{1}{10}\frac{4}{10}\frac{1}{10}3\frac{1}{2}$$

THE FIBONACCI NUMBERS

Toward the latter part of Chapter 12, nestled between problems involving the division of food and money, Leonardo throws in a whimsical problem about a growing rabbit population. He did not invent the problem; it dates back at least to the Indian mathematicians in the early centuries of the Current Era who developed the number system *Liber abbaci* describes. He clearly realized, however, as did his Hindu predecessors, that it is an excellent, easy problem for practicing how to use the new number system. And so he included it. What he could not foresee, of course, was that, although later generations of historians of mathematics would consider *Liber abbaci* one of the most influential books of all time, to most people, its author's greatest fame would rest on this one little problem.[3]

---

[3]   A Google search I carried out in September 2015, while writing this chapter, returned 588,000 hits for "Fibonacci sequence" and 607,000 for "Fibonacci numbers," but significantly fewer hits—132,000—for "Liber abaci." If these search terms don't mean anything to you, simply read on a little. Fortunately, Leonardo did not live to see how history has treated his legacy.

In what was to become his most famous passage, Leonardo wrote his way into present-day popular culture with these words:

HOW MANY PAIRS OF RABBITS ARE CREATED BY ONE PAIR IN ONE YEAR.

A certain man had one pair of rabbits together in a certain enclosed place, and one wishes to know how many are created from the pair in one year when it is the nature of them in a single month to bear another pair, and in the second month those born to bear also. [p. 404]

As usual, Leonardo explained the solution in full detail, but the modern reader can rapidly discern the solution method by glancing at the table Leonardo also presented, giving the rabbit population each month:

beginning	1
first	2
second	3
third	5
fourth	8
fifth	13
sixth	21
seventh	34
eighth	55
ninth	89
tenth	144
eleventh	233
twelth	377

The general rule is that each successive number is the result of adding together the previous two; thus, $1 + 2 = 3$, $2 + 3 = 5$, $3 + 5 = 8$, etc. As Leonardo observes at the end of his solution, although he has calculated the population at the end of one year, namely 377, this simple rule gives you the population after any number of months.

The numbers generated by the addition process Leonardo described to solve the rabbit problem are known today as the Fibonacci numbers. They first appeared, it seems, in the *Chandahshastra* (The Art of Prosody) written by the Sanskrit grammarian Pingala sometime between 450 and 200 BCE. Prosody was important in ancient Indian ritual. In the sixth century, the Indian mathematician Virahanka showed how the sequence arises in the analysis of meters with long and short syllables. Subsequently, the Jain philosopher Hemachandra (ca. 1150) composed a text about them.

The Fibonacci numbers were so named by the French mathematician Edouard Lucas in the 1870s, after the French historian Guillaume Libri gave Leonardo the nickname Fibonacci in 1838. A lot of the initial—and subsequent—fascination with them results from the surprising frequency with which they seem to arise when you go out into the garden and count things.

For example, if you count petals on flowers, you find the total is a Fibonacci number more often than you would expect from pure chance. For instance, an iris has 3 petals; primroses, buttercups, wild roses, larkspur, and columbine have 5; delphiniums have 8; ragwort, corn marigold, and cineria 13; asters, black-eyed Susan, and chicory 21; daisies 13, 21, or 34; and Michaelmas daisies 55 or 89.

For another example, if you look at a sunflower head or the base of a pinecone, you will detect spirals going in opposite directions. If you count those spirals, you will find that the sunflower has 21, 34, 55, 89, or 144 clockwise, paired respectively with 34, 55, 89, 144, or 233 counterclockwise; a pinecone has 8 clockwise spirals and 13 counterclockwise.

A third example arises in phyllotaxis, the study of the arrangement of leaves on plant stems. If you look at the way the leaves are arranged on the stem of a plant, you will find that, as you go up the stem, they spiral round. Now do some counting. Start at one leaf and let p be the number of complete turns of the spiral before you find a second leaf directly above the first. Let q be the number of leaves you encounter going from that first one to the last in the process (excluding the first one). The ratio p/q is called the divergence of the plant.

Common divergences are: elm, linden, lime, and some common grasses 1/2; beech, hazel, blackberry, sedges, and some grasses 1/3; oak, cherry, apple, holly, plum, and common groundsel 2/5; poplar, rose, pear, and willow 3/8; almonds, pussy willow, and leeks 5/13.

In none of the preceding three examples do you get Fibonacci numbers with all species, but they do occur more often than you would expect by pure chance, so it makes sense to see if there is a scientific explanation. It took several decades to figure it out, and not all the details have yet been fully determined, but there does seem to be an explanation. The key mathematical fact underlying nature's seeming preference for Fibonacci numbers is their close connection to an equally famous mathematical constant known as the *golden ratio*.

Often denoted by the Greek letter φ (phi), the golden ratio is, like that other mathematical constant π, an irrational number—a number whose decimal expansion continues forever, without ever settling into a regular, repeating pattern. The decimal expansion of π begins 3.14159; φ starts out 1.61803.

The number φ first appears in Euclid's *Elements* (written around 350 BCE), where it arises as the answer to a geometry problem about dividing up a line. Euclid gave it the name "extreme and mean ratio."

In the fifteenth century, the Italian mathematician Luca Pacioli gave it the more evocative name "divine proportion," publishing a three-volume work by that title. It acquired the alternative name "golden ratio" even more recently, in 1835, in a book written by the mathematician Martin Ohm (whose physicist brother discovered Ohm's law).

With the acquisition of not one but two suggestive names, one hinting at God, the other at wealth, it is perhaps not too surprising that various false beliefs and superstitions about the number soon appeared and began to flourish.

One such story is that the ancient Greeks believed the golden ratio is the aspect ratio of the rectangle the human eye finds the most aesthetically pleasing. Accordingly, the story continues, the Greeks incorporated it into much of their architecture, including the Parthenon in Athens, ensuring that wherever they went in their glorious cities, their eyes would be met with so-called golden (or divine) rectangles. This is clearly false on mathematical grounds. As an irrational number, the golden ratio is not a ratio of any two human measurements (which must perforce be made to a finite number of decimal places, hence rational), and thus cannot be the aspect ratio of any rectangle that can be physically measured, such as a wall of a building. On a more down-to-earth level, actual measurements of the Parthenon and other ancient buildings do not support the claim that they were built using the number.

Though there is no evidence either way whether or not the ancient Greeks felt that Euclid's "extreme and mean ratio" was even close to the most perfect proportion for a rectangle, many modern humans definitely do not. Numerous tests have failed to produce any one rectangle that most observers prefer; moreover, preferences change from one day to the next, and are easily influenced by other factors.

Claims that contemporary architects have based many of their designs on the golden ratio are in most cases without merit, though a small number did just that, including the great French architect popularly known as Le Corbusier.

Another spurious appearance of the golden ratio often cited in popular literature is that, if you measure the distance from the tip of your head to the floor, and then divide that by the distance from your navel to the floor, you will obtain the golden ratio. Again, the irrationality of the golden "ratio" makes this a theoretical impossibility. But that is the least problem with this nonsensical claim. If you measure the human body, there is a lot of variation. As it happens, the answers will always be fairly close to 1.6. But there's nothing special about 1.6. Why not say the answer is 1.603? More to the point, there's no reason to divide the human body by the navel. If you spend half an hour or so taking measurements of various parts of the body and tabulating the results, you will find any number of pairs of figures whose ratio is close to 1.6. Or 1.2. Or 1.8.

Other claims often bandied about assert a connection between the golden ratio and art. True, some artists have flirted with φ, but you have to be careful to distinguish fact from fiction. The oft repeated claims that Leonardo Da Vinci believed the golden ratio is the ratio of the height to the width of a "perfect" human face and that he used φ in his *Vitruvian Man* painting seem to be without foundation. Also without proof are the equally common claims that Boticelli used φ to proportion Venus in his famous painting *The Birth of Venus* and that Georges Seurat based his painting *The Parade of a Circus* on φ. Painters who definitely did make use of φ include Paul Sérusier, Juan Gris, and Giro Severini, all from the early nineteenth century, and Salvador Dali in the twentieth, but all four seem to have been experimenting with φ for its own sake, rather than for some intrinsic aesthetic reason.

Claims about musicians using the golden ratio in their compositions likewise seem to be without any foundation, save perhaps for some works by Debussy.

Despite the appearance of a widely available scholarly article[4] and an entire book[5] that examine all the evidence in support of the many claims made about the golden ratio and find most of that evidence sorely lacking, unsubstantiated and almost certainly false claims such as those mentioned above continue to circulate. Leonardo would turn in his grave were he to discover what is going on.

The one collection of claims about the golden ratio that do seem to hold up to close scrutiny are those pertaining to the growth of plants. In placing petals on flowers, seeds in flower heads, and leaves on plant stems, nature's inevitable preference for efficiency (this is just a fanciful way of describing the effects of natural selection) leads it to place them in a fashion that depends on the golden ratio, which turns out to have a mathematical property that results in optimized structure.

The Fibonacci numbers then appear because of that connection to the golden ratio I alluded to earlier, which was first noticed and verified in the nineteenth century. Namely, if you proceed along the Fibonacci sequence, dividing each Fibonacci number by the one that precedes it, the answers you obtain grow steadily closer to the golden ratio—in mathematical terms, the limit of those ratios is the golden ratio. (Here is how the first few values work out: $2/1 = 2$; $3/2 = 1.5$; $5/3 = 1.666$; $8/5 = 1.6$; $13/8 = 1.625$; $21/13 = 1.615$; $34/21 = 1.619$; $55/34 = 1.618$.) Since φ is an irrational number, whereas the number of petals, spirals, or stamens in any plant or flower has to be a whole number, Nature "rounds off" to the nearest whole number, and because of the above limit property, this will tend to be a Fibonacci number.

---

[4]   Markowsky, 1992.
[5]   Livio, 2002.

Knowing why the Fibonacci numbers arise in the growth of flowers and plants does not, however, imbue those numbers with other miraculous properties, such as being able to use them to predict the fluctuations in the stock market. Nevertheless, the lack of any plausible theory has not prevented the proliferation of a number of schemes claiming to do just that. If you look through the thousands of hits that come up when you carry out a Google search on "Fibonacci," you will find that a fair number of them are for investment companies that claim to achieve a better-than-the-rest performance by basing their actions on the Fibonacci sequence. Don't ask me to explain how it works; I've looked at some of those sites and their claims seem like wishful thinking to me. For some reason, otherwise sane, sensible people seem to lose their normally critical faculties when it comes to the golden ratio and the Fibonacci numbers.

Not so, I should add, the mathematicians who in 1963 formed the Fibonacci Society, which continues to publish a regular mathematical journal, the *Fibonacci Quarterly*. The society and its journal are devoted to mathematical investigations of the Fibonacci numbers, and sequences like it. For the most part, it is not deep mathematics—it generally goes by the name "recreational mathematics"—but it is all for real, and it can be an immense source of fun, since number sequences generated in a fashion like the original Fibonacci sequence have many fascinating mathematical properties. Leonardo would most definitely have approved of that part of his legacy.

FINAL WORDS

It took me several weeks to work my way through Sigler's translation. Despite my great interest in Leonardo and his work, each time I sat down to read some more, I found my mind starting to wander. In this respect, *Liber abbaci* did not have the same staying power for me as *Harry Potter and the Order of the Phoenix* (at 870 pages, considerably longer than *Liber abbaci*) did for its many millions of readers. Though it pains me to say it, Leonardo's hugely significant tome was every bit as boring as Professor Barozzi had said it was when we met in Bologna.

Well, no. Bored is not exactly right. First, it has great historical significance. It was, after all, a book that changed the world. And, as a mathematician, I was interested in the often ingenious ways Leonardo set about solving the various problems he presented.

The problem was, it was all very basic material that today we can handle much more efficiently with better methods, and with small devices we carry around in our pockets. Not only were the contents very basic, there were a lot of them—necessary for a thirteenth-century reader but overkill in the twenty-first.

I have to confess that, in the end, I never did really *read* the text from start to finish. I poked around in it, read bits here and there, skimmed over entire sections, enough to get a good feeling for the contents—and for the man who had written the words I had before me. (Mind you, I have never read a single word in any Harry Potter book. Just never felt the urge.)

There was though one thing that surprised me: *Liber abbaci*'s ending. The book ended very abruptly with these words:

> And let us say to you, I multiplied 30-fold of a census by 30, and that which resulted was equal to the sum of 30 denari and 30-fold the same census; you put the thing for the census, and you multiply the 30 things by the 30 yielding 900 things that are equal to 30 things plus 30 denari; you take away the 30 things from both parts; there will remain 870 things equal

to 30 denari; you therefore divide the 30 by the 870 yielding
1/29 denaro for the amount of the thing. [p. 615]

Leonardo is discussing an arithmetic problem solved by algebra.
"Census" is the Latin translation of the Arabic *mal* (amount of
money, or fund), and is used in many Arabic arithmetic problems,
no matter what method is used to solve it. ("Mal" and "census" are
also the names of the second degree unknown in medieval algebra.)

The book does not end with a conclusion. Leonardo gives no
reflection on what he has accomplished, no suggestion of new work
to be done, new things to try. He simply completes his description
of the solution to yet another problem and then stops writing.

Was there something more that has been lost? I wondered. Most
likely the answer is no. I discussed this point with Franci on a later
visit, and she said she felt it was completely understandable—that
he had accomplished what he set out to do, and at that point he
stopped. Another expert on medieval manuscripts I consulted
in writing *The Man of Numbers*, Professor Jeffrey Oaks of the University of Indianapolis, also informed me that many medieval
mathematical texts also end in this manner. The most you may
find is a brief statement about the copyist and the date, perhaps
with thanks to God. Authors simply did not write conclusions the
way we do today.

I am sure their explanation is probably correct for *Liber abbaci*
too, but the romantic streak in me wants to believe otherwise.
I would like to think that Leonardo knew that, with this book,
he would change the world. In which case, I think he would have
ended his book with some sort of flourish, a concluding nod toward history that has not survived to this day. Or—and now I am
being highly fanciful—perhaps the abrupt ending was his way of
indicating that there was no end—that *Liber abbaci* was merely a
beginning of much more that would follow. I like that.

Postscript: The Solution to Fibonacci's Problem of the Birds

I presented the following *Liber abbaci* problem on page 117:

> A certain man buys 30 birds which are partridges, pigeons, and sparrows, for 30 denari. A partridge he buys for 3 denari, a pigeon for 2 denari, and 2 sparrows for 1 denaro, namely 1 sparrow for ½ denaro. It is sought how many birds he buys of each kind.

Here is a modern solution using algebra. You let x be the number of partridges, y the number of pigeons, and z the number of sparrows. The information you are given then leads to two equations:

$$x + y + z = 30 \text{ (the number of birds bought equals 30)}$$
$$3x + 2y + \frac{1}{2}z = 30 \text{ (the total price paid equals 30)}$$

As I noted at the time, this looks like an impossible task, since you have three unknowns but only two equations.

But as I hinted, the problem provides you with a crucial additional piece of information that enables you to solve it: The values of the three unknowns must all be positive whole numbers. (You are told he buys three kinds of birds, so none of the unknowns can be zero, and he surely does not buy fractions of birds.)

Start by doubling every term in the second equation to get rid of that fraction:

$$x + y + z = 30$$
$$6x + 4y + z = 60$$

Subtract the first equation from the second to eliminate z:

$$5x + 3y = 30$$

Notice that 5 divides the first term and the third, so it must also divide y. So y is one of 5, 10, 15, etc. But y cannot be 10 or anything bigger, since then it could not satisfy that last equation! Thus y = 5. It follows that $x = 3$ and $z = 22$. Neat, eh?

CHAPTER 11

Manuscript Hunting, Part I (Failures)

Thanks to the labors of Laurence Sigler, the year after I returned home from my September 2002 visit to Italy I was able to read *Liber abbaci*. That enabled me to understand the book's contents, but gave me little sense of what the original must have looked like. I wondered what it would feel like to hold in my hands an early parchment manuscript of the work, one that we can presume would closely resemble the one Leonardo himself wrote? For me, it was not enough to know what Leonardo wrote. I wanted to know what an early manuscript looked like, how it felt, how it smelled. That would be the next part of my quest, when I made my next trip to Italy, later in the year. One thing was certain: Achieving that goal would require a lot more effort than placing an order on Amazon. I was to discover it would be a much more difficult task than I could have ever imagined.

The hunt started out promisingly. When I researched the matter, I found that, unlike the location of Leonardo's statue, where it was hard sorting out the fact from the fiction, the various authorities seemed to be agreed on where the oldest copies of *Liber abbaci* are kept.

As I had discovered early in my Leonardo research, there are 14 known manuscripts, all copies of Leonardo's second, 1228 version of the book, not the original 1202 edition. Seven are mere fragments, consisting of between one-and-a-half and three of the book's fifteen chapters. Of the remaining seven, three are essentially complete and are generally regarded as the most significant. Those three are in Italy.

As Professor Barozzi correctly observed when we met in Bologna, one of the complete manuscripts is in the Vatican Library in Rome, where it bears the reference mark Vatican Palatino #1343. This manuscript, from which Chapter 10 is missing, is believed to date back to the late thirteenth century. Another, also believed to date from the late thirteenth or perhaps the early fourteenth century, is in the Biblioteca Nazionale Centrale di Firenze (BNCF—Florence National Central Library), where it is listed in the catalogue as Conventi Sopressi C.1.2616. Although reports I read said that the manuscript was badly faded, and indeed "that a later hand found it necessary to rewrite what was there," this manuscript is complete. That probably explains why the publisher Baldassarre Boncompagni used it as the basis for his first printed edition in the mid-nineteenth century, even though it is not the best preserved, nor perhaps the oldest. The third one, generally believed to date from the thirteenth century as well and, according to some scholars possibly the oldest—though others have suggested it may be as much as a century later—is housed in the Biblioteca Communale di Siena (Siena Public Library).

Of the remaining, more fragmented manuscripts, four are housed in the BNCF, along with the one mentioned above; one is in the Biblioteca Laurentiana Gadd in Florence (Gadd. Reliqui 36, dated to the fourteenth century); one is in the Biblioteca Riccardiana in Florence; one in the Biblioteca Ambrosiana in Milan; one in the Biblioteca Nazionale Centrale in Naples; and three in Paris (one in the Bibliothèque Mazarine, two in the Bibliothèque National de France).[1]

At first, I was surprised to learn that none of the early manuscripts were in Pisa, Leonardo's home town. But then it dawned on me. In the days when every copy of a manuscript had to be prepared laboriously by hand, there would be little incentive to keep a copy in a city that already had a manuscript. Pisa, of course, had Leonardo's own original. Thus, any copies made would most likely have been for use elsewhere. Rome, Florence, and Siena would be obvious places where copies would be found. Whether or not my theory is correct, it leaves open the tantalizing thought that maybe somewhere in Pisa, deep in the cellar of some church or monastery, lies Leonardo's own original manuscript of *Liber abbaci*.

The copy in Siena, having reference number L.IV.20, was the manuscript that Raffaella Franci suggested I should consult, when we first met in the late summer of 2002. It was the manuscript on which she had based much of her own work. Although it is missing much of Chapter 15, by all accounts it is the one that would best meet my particular desire to examine the artifact rather than its contents. For one thing, it is generally regarded as the most attractive and best preserved of all the *Liber abbaci* manuscripts. Leonardo scholar Professor R. E. Grimm, the man who critiqued the widely circulated translation of the autobiography in the prologue to *Liber abbaci*, has described it as "the best existing manuscript." Franci showed me photographs of some of the more colorful pages from the manuscript, displaying a clear, elegant script and beautiful hand-drawn graphics and illustrations. But to me, the most significant feature of all is that it is believed by many to have been completed around 1275, just a few years after Leonardo

---

[1]   For more details, see Hughes, 2004.

would have died. The scribe who wrote it may have started work during Leonardo's lifetime. Perhaps he had even met Leonardo, or shown him some of the early pages.

I had asked Franci what preparations I needed to make in order to examine this manuscript. What kind of permissions would I need to gain access to such a rare object? "None whatsoever," she replied, surprised. "It's just a manuscript in the local public library. Anyone can see it." All I would have to do, it seemed, was walk into the library, ask to see manuscript number L.IV.20, and then examine it in the reference room. I would complete my study in view of the reference librarian, who would bring it to me from the stacks, but I could examine it for as long as I wanted.

Having spent most of my life as a scholar in England and the United States, I was used to rare manuscripts being kept in special collections in academic libraries. In general, only a bona fide scholar could examine one, and then only after a long process of obtaining permissions. The idea that I—or anyone else—could simply walk in off the streets into a public library and lay my hands on such a treasure was almost inconceivable. But that's how it is in Italy.

I was so excited by what she told me that I wished I could have examined it immediately, but my remaining time in Siena had been limited. Besides, I knew I would be in Italy again the following year, so I decided to wait until I would have more time at my disposal for such an important occasion. The intervening months would give me more time to learn about the manuscript, following the leads Franci had given me. I would return the following August. And so I did.

THE SIENA MANUSCRIPT—ALMOST

During the summer of 2003, the whole of Europe was suffering under a record-breaking heat wave. I spent the first two days of my weeklong stay in Italy in Pisa, taking more photographs of the

Leonardo statue and the places I had seen on my visit the previous year, including the two Fibonacci street names. Then I drove down to Siena—this time I had a rental car—traveling by way of Livorno, the large port city that today occupies the site of the Porta Pisano, from where Leonardo had set sail for Bugia. Early the next morning, in the sweltering heat, I made my way through the crowds of tourists thronging the narrow medieval streets of Siena, to make my first acquaintance with *the oldest existing manuscript of a book that changed the course of Western civilization.*

[I am afraid you'll have to endure my repeating this mantra so often, but throughout my work on the project I was keenly aware of the enormous role in human advancement Leonardo's book played, no more so than when I was searching for, and in due course holding in my hands, some of the earliest known manuscript copies.]

By chance, I was in Siena during the annual Palio, a weeklong historical festival that culminates in a furious horse race, where young men from ten of the seventeen Contadas (districts) that make up the city ride bareback round the city's central square, the Piazza del Campo, to the enthusiastic cheers of a crowd that packs into the square to enjoy the spectacle. The entire city was bedecked in colorful flags. I resolved to try to see the race when it took place on the Saturday, two days later. But first, I wanted to examine manuscript L.IV.20.

The Biblioteca Communale di Siena is at number 5 Via della Sapienza (Street of Knowledge), a few minutes' walk from the Piazza del Campo. From the outside it is a large, grey, stone building, darkened by age, dating from the middle of the eighteenth century. Large, modern glass doors set within the stone-arched entrance gave the impression more of a grand nineteenth-century European hotel that has been modernized, rather than a public library. But then, as I had already learned, the treasures that this old building

FIGURE 15. The entrance to the Siena Public Library, when I visited in 2003.

held—of which *Liber abbaci* was just one of many—made this quite unlike any library I had been in.

[When I returned to Siena in 2009, I discovered that the original entrance had been closed off, and entry into the library was now through another, rather nondescript doorway (number 3) farther down the street. To gain entry, you now have to pass through a security barrier. Altogether much less attractive, but if it keeps Leonardo's manuscript safer, that's fine by me.]

The sense of surrealness I felt continued as I left the noise, bustle, and heat of the outside street and stepped into the cool, quiet interior. My first impression was that I had wandered onto the set of a Hollywood movie intended to portray a long forgotten, dusty old library full of literary treasures—perhaps for a scene in one of those Harrison Ford *Indiana Jones* movies. On all sides, the walls were stacked high with faded old books, stretching all the way up to the high vaulted ceiling. Metal spiral staircases led to the metal gantry that circled the entire building, giving access to the higher books. The building had an overpowering musty smell of old paper. The visual impact of so many ancient books at first detracted from the second striking feature of this place. I expected to see the librarian seated at her desk, and the old wooden tables at which men and women sat reading books and making notes. But the circular arrays of computers I saw before me seemed completely at odds with everything else, an unexpected intrusion of the twenty-first Century into the eighteenth. Even a library as old as this, it seemed, could benefit from online cataloguing systems and search engines.

For all their obvious age, however, the books I could see in this room were not the valuable, rare manuscripts of the library's collection. To gain access to those, I had to register with the woman at the front desk, and inform her of my interest. She appeared to speak no English, so I continued with the Italian I had begun with—poor and broken, but generally adequate to make my request known. After a few moments, she beckoned me to follow her through a door into another room, clearly the reference reading room. There, an earnest young man gave me a card to fill out, on which I had to give the details of the manuscript I had come to see. "Un momento," he said, gesturing to a seat at which I should sit—he too seemed to speak no English— and disappeared behind a large wooden door that led into what I assumed was the room where the historical collections were kept. I could scarcely contain my excitement. Within a few minutes,

I would be holding what was possibly the oldest manuscript of *Liber abbaci* in existence.

The man had left the door slightly ajar, and I could see him walking back and forth, clutching my card in his hand, a puzzled look on his face. After a few moments, he stepped outside, gestured to me to remain where I was, and left the reading room. Eventually he came back. "The manuscript is not here," he said in Italian. "It is in Germany." He seemed as surprised as I was.

In fact, I was more than surprised; I was stunned. The manuscript had been here for centuries. It is extremely rare for any library to loan out any rare item; scholars who wish to consult a valuable manuscript normally travel to the library where it is housed, as I had done. Where exactly in Germany was it? Why had it been moved there? Was it merely on loan, or had it been permanently transferred? When would it be returned to Siena? I struggled to find the Italian to ask these questions, but the man said he had no idea. All he knew was that the records said that manuscript L.IV.20 was presently in Germany.

My Italian was not sufficient to take my questions much further, and besides, I got the impression that there was no one around at the time who would know the answers. I would have to give up on the quest for the day and write to the chief librarian upon my return to the United States. I hoped that a rapid email exchange would secure the information I wanted, and I would have to make plans to examine the manuscript at a later date, either in Germany or back in Siena.

The man could see I was disappointed and shrugged in a good-natured way, as I collected my notepad and pen and turned to leave. I had no doubt I would get to see the manuscript eventually, but clearly it would not be on this visit to Europe. My next opportunity would be the fall, when I was scheduled to return to Italy to receive the Peano Prize.[2]

---

[2] For the Italian editions of my books *The Math Gene* and *The Language of Mathematics*.

On the way out, I decided that, to ease my disappointment, I would drive up to Florence the next morning and see the copy of *Liber abbaci* in the Biblioteca Nazionale Centrale di Firenze. If I could see the Florence manuscript now, and the Siena one on my next visit, I would feel that my trip had not been in vain. Later that evening I checked on the BNSF website and learned that the library was open on weekdays between the hours of 8:15 in the morning and 7:00 in the evening.

CHIUSO

Traffic was unexpectedly light as I sped up the Siena–Firenze Autostrada early the next morning, the Friday. It was also strangely light in Florence itself, and to my surprise I had no difficulty finding a parking space within a short walk of the Piazza dei Cavalleggeri, the small square on the north bank of the Arno where the BNCF is located, a short distance from the famous Uffizi Museum.

The imposing main entrance to the BNCF was surrounded by a large wooden fence, over which a large printed sign explained that the building was currently undergoing extensive renovations, during which time entrance could be gained through a side door.

I had no difficulty finding the alternative entrance, but to my surprise the gate was locked. The guard house in the yard behind the gate was empty and the entire place seemed deserted. I rang the bell, but no one appeared. I looked for a notice explaining what was going on. But neither in front of the main entrance nor at the side doorway was there any kind of information sheet, not even one giving the library's normal opening hours, let alone information about today's closure.

I was reminded of the Sunday a year earlier when I had taken a bus ride from Siena to the nearby medieval hilltop town of San Gimignano, whose many tall towers make it one of Italy's most

FIGURE 16. The Florence National Library, which was closed when I tried to visit in 2003.

popular tourist attractions. Before I had left Siena, I had checked the bus schedule carefully to determine when the last bus back to Siena departed from San Gimignano, and when I got off the bus in San Gimignano I checked again. Not wanting to end up having to spend the night there, I made sure I was back at the bus stop a good ten minutes before the scheduled departure time. So too did a dozen or so other tourists—a mixed group of mostly young people, including Germans, French, Scandinavians, Japanese, British, and Americans. When half an hour had passed and still no bus had arrived, we began to talk to each other, wondering what had happened. Another quarter of an hour passed, and still there was no bus. It began to look like it was going to rain. I decided to look very carefully at the bus timetable posted behind glass in a wooden framed case on the rear wall of the bus shelter. There, in the lower left-hand corner, almost impossible to read beneath the wooden frame, it gave the date on which the timetable had been published. It was from 1999. Clearly, at some stage in the intervening three years, the bus company had changed the schedule, but had not bothered to post a new timetable. "This is Italy," we all shrugged. A short while later, a local bus turned up that was heading to Poggibonsi, a nearby town that several of us knew was on the Florence to Siena railway line, so we all piled aboard, the younger tourists filling the aisles with their bulky rucksacks so that it was virtually impossible to move along the bus. As it turned out, we did not have to complete our journey by train, since another local bus went from Poggibonsi to Siena. But the episode reminded me of what I had heard from others about the Italians' extremely laid-back approach to life.

Now, a year later, here I was again, victim to that same attitude. And just like that first time in San Gimignano, I felt none of the frustration or anger I would have done if I had driven 50 miles to a national library in the United States and found it closed, with no opening hours posted. After all, it was that very relaxed Italian approach to life that would—eventually, I knew—allow me to hold in my hands two of the most valuable scientific historical documents in the world. Having to return at a later date was a relatively small price to pay.

I eventually learned, from talking with some British tourists, that the day I had chosen to visit the library in Florence, August 15, was the biggest annual national holiday in Italy, and everything was "chiuso"—closed. The August 15 holiday is so well established in Italy that no one who lives there would see any need to post a notice about the closure of a library—or anything else for that matter—on that day.

CONSOLATION PRIZE: A DETOUR TO VIA FIBONACCI

And so, my first two attempts to examine early copies of *Liber abbaci* had failed. I made use of the remainder of the morning heading out to the northern part of Florence, where there is a short, sleepy, domestic street named Via Leonardo Fibonacci. At the very least I could leave Italy with some photographs of the street name! In this, at least, I would be rewarded handsomely. Although the entire street is only a hundred meters or so in length, it boasts no fewer than 11 street signs, each bearing the name "Fibonacci," which must make it one of the most identified streets in the world.

Two of the signs are carved in polished marble affixed to the wall, three have blue lettering painted on white metal, also attached to the wall, and the remainder are modern white metal signs with black lettering, mounted on metal poles. I took a photograph of each one.

Two of the signs are next to (what at the time was) a small garage built into the building, housing a small motor scooter, a workbench, and a few tools. The elderly man who was working inside rushed out when he saw me taking photographs, clearly disturbed by my actions. He seemed not to believe me when I told him that his street was named after a famous mathematician and that I was a mathematician who was photographing the street signs, not the interior of his garage. I assume he feared I was from the local council, perhaps checking up on a complaint from a neighbor. He went back inside, but kept watching me suspiciously until I moved away to photograph the signs further down the street.

Of particular interest to me was the fact that house number 12 on the Via Leonardo Fibonacci is the home of the Italian Computer Society's Tuscany branch. (The nameplate on the gate reads Associazione Informatici Professionisti: Italian Computer Society—Sede Toscana.) Continuing the scientific theme, house number 1A, at

FIGURE 17. Via Fibonacci in Florence. This short street boasts no fewer than 11 Fibonacci signs. I photographed all of them in 2004.

the corner of the square at the southern end of the street, is home to the Instituti di Neuroscienze (INS).

I may not have been able to examine two hugely influential medieval manuscripts on this research trip, but I was at least able to return home with the satisfaction of having taken some excellent photographs of street signs in one of the less-tourist-frequented parts of Florence.

CHAPTER 12

Manuscript Hunting, Part II (Success at Last)

I did not have time to visit Siena or Florence when I flew out in October 2003 to receive the Peano Prize, but I was in Tuscany again the following spring—my fourth trip to Italy in a two-year period. This time my visit was primarily for a vacation, but while there I intended to try once again to see at least the Siena Public Library copy of *Liber abbaci* (I was staying in a vacation rental just south of Siena), and possibly the Florence manuscript as well.

On May 13, shortly after noon, I entered the Siena Public Library and strode up to the front desk. "Vorrei vedere un manoscritto," I said. ("I would like to view a manuscript.") The young lady at the desk replied in Italian far too rapid for me to understand, but I caught the gist. The reference library was closed for the long Italian lunch break, and would re-open at 3:30 PM. I would have to come back in three hours' time. Here we go again, I thought.

Three hours—and several cappuccinos at an outdoor café on the edge of the Piazza del Campo—later, I was back in the library, repeating my request to consult a manuscript. The young lady at the desk had been replaced by two elderly gentlemen, one of whom escorted me to the manuscript room at the back of the library, where my search had ended unsuccessfully the year before. There, a very friendly young man gave me a form to fill in, requesting the manuscript I wanted to see. "Please, take a seat," he said to me in Italian, gesturing toward one of the study tables. All around me, scholars were earnestly poring over ancient texts. Would I get that far on this, my second attempt? As had happened the previous year, I could see the assistant peering at the shelves in the manuscript room, searching for the volume I wanted. The moment seemed to drag on eternally, but then, I saw the young man step forward and pull from the shelves (which I could not see) a thick, old, brown volume. He brought it over and placed it on the desk before me.

There it was. The moment had arrived. I held in my hands what was possibly the world's oldest known copy of *Liber abbaci*. The book that changed the world. A manuscript that may well have been started while Leonardo was still alive.

I find it hard to convey the feelings I had at that moment. I have never felt the sense of reverence for major movie stars or famous musicians or singers that many people seem to have, but I imagine that the sensation I had when I first held the Siena manuscript of *Liber abbaci* in my hands was not unlike the feeling many people say they have when they finally meet that favorite celebrity.

The book was large and heavy. The thick cover was a discolored brown, with gold lettering. The spine bore the inscription

LION. PISANI
DE ABACO

near the top, and the Siena Public Library reference number

L.IV.20

near the bottom. The pages of clean, thick white paper that formed the inner binding made it clear that the volume had been re-bound fairly recently. The inside of the back cover carried a small stick-on label that bore the legend

Guiseppe Masi
Restauratore
Firenze

The manuscript, which is missing much of the final chapter, comprised 224 sheets, each one written on both sides. The page numbers—running from 1 to 224, appropriately in Hindu-Arabic numerals—had been added later, one on the top center of the front side of each sheet. The parchment was thick and stiff but not brittle. I took out a small ruler I had brought for the purpose. Each sheet measured 20.5 cm wide by 30 cm high. Discounting the thick covers, the manuscript was 5 cm thick. Apart from the front page, which had partly disintegrated and had been attached to a backing sheet, each page was in remarkably good condition. An occasional page had a hole in it, and the outside edges of some pages had worn away under the cumulative effects of 800 years worth of page-turning hands.

Many pages were a whitish-cream color, others a light brown, and many had discolored. Each page had been carefully ruled with a grid to guide the lettering, and the text was written in brown ink, with every numeral in red. At the start of some paragraphs, the initial letter was enlarged and also in red. In addition, the scribe had embellished some chapter openings with large stylized letters in blue and gold.

Despite the damage, the first page still showed the original title near the top:

Aritmetica Leonardi Bigholli de Pisa

There was that mysterious name "Bigholli" I had come across a number of times. What was the origin of this name? In his classic text *History of Mathematics*, D. E. Smith writes:

> Possibly it was his [Fibonacci's] indulgence in travel that caused him to write his name occasionally as Leonardo Bigollo, since in Tuscany bigollo meant a traveler. The word also means blockhead, and it has been thought that he had been so called by the professors of his day because he was not a product of their schools, and that he retaliated by adopting the name simply to show the learned world what a blockhead could do.[1]

Somehow, I just could not swallow the blockhead story. Leonardo was trying to change the world, not get back at his teachers. My helpful Italian correspondent, the mathematics teacher Gian Marco Rinaldi, had sent me an email commenting on the name. People in Tuscany do not use the word nowadays, he informed me, and Italian dictionaries do not list it. Both in Tuscan dialect and literary Italian there is a similar sounding verb *bighellare*, in modern usage *bighellonare*, which means to wander aimlessly, and a *bighellone* is a man who squanders his time and does nothing useful. But this would be such an inappropriate description of Leonardo that I think Sgr. Rinaldi is right when he suggests that Leonardo's term *bigollo* is unrelated to the modern *bighellone*.

Another reason to doubt any depreciative connotation to *bigollo* comes from a declaration by the Comune (the town government, at that time an independent state or republic) of Pisa in 1241, recognizing Leonardo's great contributions and awarding an amount of money to be given to him each year. The text of the declaration refers to him as Leonardi Bigolli. Rodolfo Bernardini,

[1]   Smith, 1951, p. 216.

a local scholar who has studied the issue, advances yet another proposal for the meaning. He says that *Bigollo* might indicate a man who knows two languages (in Leonardo's case, Latin and Arabic), with reference to the medieval Latin word *biglosus*, which has precisely that meaning. In any event, the uncertainty surrounding the origin of the very name used to refer to Leonardo is yet another mystery about the man.

Turning to the final page of the manuscript, I saw that it ended with (as far as I could make out) the legend

Exptir cbet aritmetica Leonardo bigholli de pisa

just squeezed in at the bottom of the page.

I spent about an hour going through the entire manuscript, turning the pages carefully. I don't really read Latin, but as a mathematician with a few years of high school Latin in my past, I had little trouble understanding what was being said. It was an amazing experience. I wondered if Leonardo himself had spoken with the scribe who wrote it, or had seen the earlier leaves of this particular copy, which had been completed before his death. Who had learned arithmetic from it in the 800 years of its existence? Which famous hands had turned the very pages I was now turning? Galileo? Pacoli? Da Vinci? With so few copies available at the time, there was little doubt that I was holding one of the original intellectual bridges from the Indian and Arabic mathematics of the first millennium to modern, western European commerce, trade, and science.

As I looked through the manuscript, I made a note of the pages of which I wanted digital copies. Since the leaves were numbered only on the front side, I invented my own annotation of "a" for the front side of each sheet and "b" for the reverse side, and hoped my Italian would be up to explaining the notation to the assistant. (It turns out that this notation is in fact in common use among archivists.) I selected 12 sides in all: 1a, 1b, 3a, 5a, 5b, 24a, 31b, 53a, 53b, 87b, 153a, 207a. The assistant at the desk was very helpful. He took me into an adjacent room where I spoke with the man who would make the images and burn them onto a CD for me. "Come back the day after tomorrow (the Saturday) in the morning and they will be ready," the technician said in Italian.

When I returned on the Saturday morning to collect (and pay for) my CD, both the librarian and the technician greeted me warmly as soon as I walked in. The technician took me over to his computer to show me the images, and to confirm that they were the ones I wanted. The pages looked much cleaner and brighter on a computer screen than when I held the book itself. I thanked him and prepared to leave, but I could not resist taking another, brief look at the manuscript.

The assistant duly obliged, and once again I had the manuscript in my hands. I turned the pages once again. Although the words I was reading were not written in Leonardo's own hand, the copy was completed soon after he had died. So it was possible that the scribe who had made the copy did so directly from the original. Perhaps the two had even met and spoken about the contents. When I finally handed back the volume to the young man at the desk and left the library, I knew that I had come as close to Leonardo the man as anyone possibly could.

THE FLORENCE MANUSCRIPT

Buoyed by my success in seeing the *Liber abbaci* manuscript in Siena, a few days later, on May 20, I traveled to Florence to try once again to examine the copy in the BNCF. It was a Thursday, and this time not a national holiday, which meant that I had to negotiate the famously frantic Florentine traffic and spend considerable time looking for a place to park, eventually resorting to the method I had seen Italian colleagues using, namely, parking illegally in a residential area close to the library and accepting the inevitable fine as a parking fee. (It turned out to be EUR 33.90, which I regarded as an excellent value for finally getting my hands on *Conventi Soppressi* C.1.2616.)

The construction that had been under way on my previous visit to the library the year before had been completed, and the wooden fencing removed, leaving the library's imposing main entrance unobstructed. I strode in and entered what was clearly a security booth, through which everyone who wanted to enter the library must pass. A bored-looking woman sat behind a counter. I told her I wanted to view an ancient manuscript. Apparently my Italian was not fully up to the task, or else (and I suspect this is the more likely) the bored woman was not really listening, having pegged me as a foreign tourist who had wandered into the wrong building (probably looking for Michelangelo's *David*), since she told me that this was a reference library for scholarly research only and hence I could not enter. I told her I was a professor and wanted to see a particular ancient manuscript. She hesitated for a moment, eying my jeans and t-shirt with suspicion, clearly not her expectation of how a university professor should dress. But then she presumably decided that the easiest way to get rid of me was to allow me to enter and become someone else's problem.

So, I exchanged my passport for a visitor's day pass to the library, and the lady directed me to the information office at the back of the large, marble-floored entrance hall. From there I was directed to the first floor (second floor in US terminology), where the Rare Manuscripts section was located. At the first desk I walked up to, the assistant spoke excellent English, and directed me to a colleague in the next room. She also spoke English fluently, asking to see my library card. "Oh dear," she said, "this is not the right card to examine manuscripts. Please go to the lady at the end, who will be able to help you."

So far so good. I was now in a large reading room with a high, wood paneled ceiling, surrounded by shelves full of ancient books. Unfortunately, the lady to whom I had been directed spoke almost no English. She explained to me in Italian that in order to give me permission to examine a manuscript, I would need a letter of introduction. "I am a professor from America," I pleaded in Italian. (Fortunately, at least half the scholars seated at the two rows of reference desks that filled the room were dressed just like me, so clearly I was now past the stage of having to look like someone's erroneous expectation of a professor.) "Do you have your . . ." she replied in broken English, struggling to think of the word. "University ID," I completed for her. "Yes, here it is." I took out my wallet and removed my Stanford identity card, placing it on the desk in front of her. That did the trick, and she started to fill in a green card to give me permission to gain access to the manuscripts. "Please, I need your passport," she said. I told her I had given that up at the front desk, in exchange for my library visitor's card. "But I have to have your passport to arrange for your permission card," she explained. We were at an impasse.

At that point, the librarian asked me to wait a moment, and went through the door behind her, returning a moment later with a colleague. "Can I help?" the second lady said, in English, with an accent very similar to my own North of England Yorkshire. I explained the problem with the passport. "We'll have to get you a proper library card," she said. The visitor's card you have isn't good for what you want to do."

And so Jeannie Nesi (née Bettaney) from Manchester, who had first come to Italy 30 years earlier as a nanny and was now a librarian in the Rare Manuscripts section of the BNCF, took me under her wing and guided me, first through the 20-minute process of exchanging my visitor's pass for a regular scholar's pass (back downstairs) and then helping me to fill out the request form for the manuscript I wished to examine (upstairs again). "I'll be back in a few minutes with your book," she said. It was now three o'clock. It had taken an hour, but I was almost there.

And then I really was there. I exchanged my passport (now retrieved for me by my Manchester helper) for the thick book with a tan-colored, leather cover she had brought in to the reading room. (She explained that because of past thefts of rare manuscripts, they hand over books only in exchange for a passport or ID card, and seemed surprised when I told her how much more informal they were at the public library in Siena.)

The volume was surprisingly thinner than the one in Siena, although about the same height and width. It fastened with two brass clips. There was no writing on the outside cover. I unfastened the clasps and took out my ruler to measure the manuscript. Each page was 20 cm wide by 30 cm high, the same size as the one in Siena. Discounting the cover, it measured 4 cm thick, a full centimeter less than L.IV.20. There were 213 pages, each written on both sides, giving 426 sides altogether—almost the same number as the Siena copy.

On the inside of the front binding were pasted two tiny fractions of manuscript, clearly all that was left of the original cover page. All I could make out, and then not with total certainty, was:

Leonardi Pisani Algorism A[ritm]etica

The first page of the manuscript itself bore the following legend at the top of the page:

A C Leonardus pisanus Algorisma & Geometrie est Abbacie florenty

At least, that's what I thought it said. The writing was badly faded.

I started to turn the pages. I felt the same tingle of excitement and awe I had experienced in the Siena library, again wondering which famous scientists and mathematicians had used this very manuscript to learn arithmetic.

References to the Florence manuscript that I had seen in books and articles and on websites generally referred to it as "badly faded," but this was definitely not my first impression. On the contrary, the manuscript seemed to be in much better condition than the one in Siena. The paper was dry but not brittle, and felt slightly thinner than that of the earlier Siena copy—which explained why the bound volume was much thinner, even though the page count was almost the same. None of the pages had holes or worn edges, as is the case with L.IV.20. The basic color scheme was the same, with the Latin text in a brown-black ink and all numerals in red. The scribe who had written it certainly had the greater artistic flair of the two, decorating the margins of many pages with fancy swirls, and making much more extensive use of fancy red letters to begin paragraphs and large red, blue, and gold fancy letters to start new sections. Certainly, some of the pages were faded, some of them so badly that it was hard to discern the text—at least for someone like myself who does not know much Latin. And the paper of many pages had turned brown or grey and become spotty. On some pages I could see the guidelines the scribe had used to line up the text. But all in all, it was in great shape, given its age.

Jeannie, my Manchester guardian, was now sitting at a desk alongside the woman who issued the passes (and who was now holding my passport as ransom for the manuscript), and I asked her how I could order digital images of some of the pages. She gave me a form, and I selected the pages I wanted to have copied. She helped me by telling me the correct way to stipulate manuscript pages: a letter "r" for *recto* to refer to the front side and "v" for *verso* for the reverse side.[2] (Confusingly, the Italian word *retro* actually means back side, and *verso* means front, among other things, so I paid particular attention in completing the request form.) I ordered 1r, 1v, 4r, 14v, 20v, 27v. According to the young

---

[2]   As I indicated previously, I subsequently learned that my earlier use of an *a, b* designation is also in common use.

man at the desk where I submitted my request, the images would be saved on a CD and sent to me at my Stanford address in about two weeks' time.

And that was that. After two years of frustrated attempts, within the space of a few days I had held in my hands and examined two of the three earliest known manuscripts of *Liber abbaci*. The one remaining manuscript I wanted to examine was the one in the Vatican Library in Rome. But for now I had a vacation to enjoy. Vatican Palatino #1343 would have to await a further trip to Italy. Though the completion of this episode will take me ahead of my story, I'll tell you how it ends.

THE VATICAN MANUSCRIPT

A series of other professional commitments meant I had to put aside the Leonardo project for several years, and it was the spring of 2009 when I was finally able to resume my quest. I had just begun to plan a trip to Italy when by good fortune I received an invitation to give a lecture at a conference at the University of Verona that October. My schedule at Stanford seemed to offer me the possibility of extending my stay by a few days, so I began to work out the details of my trip. In addition to visiting Professor Franci in Siena, I wanted to go to Rome to examine the Vatican manuscript. Mindful of the trouble I had encountered on previous manuscript hunts, I went onto the Vatican Library's website to make sure that I would visit it on a day when it was open. But instead of finding a list of opening hours and days when the library would be closed, I was greeted with the following announcement (I was on the website's English version):

UNAVAILABLE COLLECTIONS

At present, all of the Library's collections are unavailable for consultation.

The Vatican Library is closed to the public since July 14, 2007
for important refurbishment works which are expected to
last for three years. The refurbishment could not be delayed
because of serious structural problems in a wing of the
sixteenth-century building which houses the Library. Tech-
nicians have established that it is impossible to reinforce
the floors and other structures without closing the Library
to the public. The management of the Library had sought to
avoid inconvenience for the public, e.g. by moving more than
300,000 volumes out of the areas where the floors had begun
to weaken; but the experts have decided that the refurbishment works could not be postponed. Once the necessity of
the structural work had been established, it was decided to
take this opportunity to modernize certain parts of the building, including new air conditioners and elevators. Likewise,
a general reorganization of the Library's collections will also
be undertaken. During the closure, many of the Library's
books will be moved to temporary stacks. However, copies
of manuscripts and of rare books are available through the
photographic reproduction service, which will continue to
operate.

So that was that. Either I waited another year—likely longer given
the inevitable delays in construction projects—or else I brought
my quest to a close without an examination of Palatino #1343.
After some thought, I decided to pass on a trip to Rome. The
Vatican manuscript would have been a nice act of closure for
me personally—and one day I expect I will look it up, should the
opportunity arise—but it would not have added to my understanding of Leonardo. My story was already as complete as I could
possibly make it. Besides, Leonardo did not "finish" *Liber abbaci*.
He simply reached a point where he stopped writing. I decided
to do the same, and resolved to complete my research by the end

of the year and complete the writing of *The Man of Numbers* the following spring. But now I have jumped well ahead in my story. I had one more Leonardo memorial to check out before my 2004 visit came to an end.

THE PROCLAMATION

In his fall 2003 email telling me the history of Leonardo's statue, Sgr. Renaldi, the mathematics teacher from near Pisa, had told me of another Fibonacci monument. I had resolved to locate it on my next visit to Pisa, in 2004.

It was a short walk from the Piazza XX Settembre, the square from where Leonardo's statue had looked out across the Ponte di Mezzo from 1926 until the end of the Second World War. After having my photograph taken on the very spot where Leonardo's (fictitious) effigy had stood—something I had wanted to do ever since Sgr. Renaldi had sent me that remarkable wartime photograph—I crossed the bridge and headed east along the Lungarno Mediceo, the road that follows the northern bank of the river.

A short walk of some 500 meters brought me to house number 30, an imposing, three-story building with attractively decorated grey stone walls. This was the Palazzo Toscanelli. It originally belonged to the Lanfranchi family, but then passed on to Alessandro della Gherardesca, who made a number of architectural modifications. In 1821–22, it was the home of Lord Byron and his circle of friends, during the period in which Pisa was called the "Paradise of Exiles." From here Lord Byron departed for Greece, where he met his death.

Today the palace hosts the Archivio di Stato (State Archives) of Pisa. The item I had come to look at was a stone tablet on a wall in the entrance hall. It was placed there on June 16, 1867. The inscription on the stone is several lines long. Following an introductory declaration written in 1865, it reproduces the text of a document of 1241, by which the Comune of Pisa (the town government—at that time Pisa was an

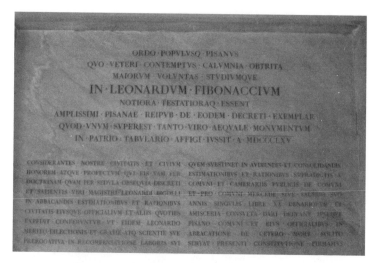

FIGURE 18. The Leonardo commemorative tablet in the State Archive in Pisa, photographed in 2004.

FIGURE 19. The commemorative tablet in situ, photographed in 2004.

FIGURE 20. The frontage of the State Archive in Pisa, photographed in 2009.

independent state or republic) decreed that Leonardo should receive an amount of money annually for his merits. (This proclamation is the only indication that Leonardo was still living in 1241.)

I took several photographs of the tablet, making sure that I managed to get a clear image of the entire text. When I was working on the final draft of *The Man of Numbers* in 2010, I tried to translate the text from the photographs. With the aid of a Latin-English dictionary, I can usually figure out Latin text, but this one was written in a stylized nineteenth-century form of the language, using words not in my dictionary, and some parts completely stumped me. I asked mathematics historian Barnabas Hughes to translate it for me, which he kindly did. He remarked that the Latin is very formal. In his translation, Hughes kept to the original Latin word order, which is different from that of English. In English, the tablet reads:

The Rulers and People of Pisa in the year 1865 after ignoring old crushing falsehoods and where the will of the Elders was to study what was better known and proven about Leonard Fibonacci ordered the city archives to file a copy of the decree by the same Most Eminent Republic of Pisa that one monument equal to so great a man survive.

[THE 1241 DECREE]

In consideration of the honor brought to the city and its citizens and their betterment by the teaching and zealous cooperation of that discreet and wise man, Master Leonardo Bigolli, as well as by his regular patriotic efforts in civic and patriotic affairs, the Pisan Commune and its Officials in certain right and conscious of our prerogative to make recompense for work that he performed in heeding and consolidating the efforts and affairs already mentioned confer upon this same Leonardo so meritorious of our love and appreciation an annual salary or reward from the Commune of 20 free denarii and the usual accompaniments. This we affirm with the present statement.

I was unable to find any information about the installation of the tablet. I assumed it was a result of the growing awareness of Leonardo's accomplishments during the nineteenth century, following Cossali's discovery of his legacy at the end of the eighteenth. Perhaps a local historian was prompted to search through the city archive, eventually finding a copy of the decree. In any event, it is the only official record ever discovered of one of the greatest men of all time. (I know. I really can't help myself.) By transcribing its words onto the memorial tablet, the City of Pisa created a public record in marble that would withstand the passage of time.

CHAPTER 13

The Missing Link

With the bulk of my archival and location research completed by the end of 2004, I did not really need to make any more trips to Italy in order to complete my project and write the Leonardo book I set out to produce. But, having fallen in love with Italy, and Tuscany in particular, during my very first visit to Siena in 1984, I continued to look for opportunities to go.

With Italian translations of my popular mathematics books selling well in Italy, invitations kept coming, so I traveled to Italy annually for several years, and on one occasion twice in the same year. In years when no invitation came, I decided to go anyway, primarily for recreation. When a knee injury in 2002 forced me to give up a 25-year daily distance running habit and take up cycling as an alternative way to get my outdoor exercise, I started to take cycling vacations in Tuscany. On each visit, I would call in to see Franci or spend a few hours in a library archive, or maybe check on a particular location I had mentioned in the by-then-completed early draft of *The Man of Numbers*.

When that book came out in 2011, I still had not seen two important Leonardo artifacts firsthand. One was the complete *Liber abbaci* manuscript in the Vatican Library. The other was the abbacus book that Franci had studied, and identified as almost certainly an early copy of Leonardo's otherwise lost "book for merchants," what I had referred to in my book as "the missing link."

Seeing either of those volumes would not broaden my understanding of Leonardo and his works. I had read Sigler's English translation of *Liber abbaci*, so I knew what Leonardo had written. But there is something visceral about holding in your hands an actual artifact that cannot be duplicated by looking at high-resolution images. So it was important to me to hold, and examine, at least one of those three complete manuscripts, Having done that with two of them, a third would make little difference.

Had Leonardo's original 1228 manuscript survived, I would have worked my network of friends and colleagues to the full for an opportunity to get as close to it as I could, and ideally hold it in my hands. To feel the texture and take in its aroma, knowing that the object I held was created by Leonardo himself. But that was not to be.

[Examining the Siena manuscript and the Florence manuscript was not quite the same as touching the original, of course. But since both were written shortly after Leonardo's lifetime, and possibly started while he was still alive, they had been close enough to have a powerful emotional impact on me. For, almost certainly, while the scribe was inking the manuscript I was holding, alongside it would have been Leonardo's original.]

With the missing-link manuscript Franci had studied, however, it was much less certain that this had been copied directly from Leonardo's own "book for merchants." The abbacus books were much shorter than *Liber abbaci*, making copying easier, and the accessibility and immediate utility of their contents made them a must-have resource for many people. As a result, copies would surely have proliferated at an increasing rate, and the vast majority would have been copies of copies, or copies of copies of copies, and so on.

True, the Franci manuscript was written not long after Leonardo's time, but long enough that at the very least it was likely a copy of a copy. In any event, that was my reason for not making an attempt to examine the missing link before. I had studied at length Franci's paper about its contents, and that was all I needed to complete *The Man of Numbers*.

It was only when my book had been published, and I started to give talks about it, that I wished I had made the effort to gain access to that manuscript. Not least because doing so would have enabled me to select pages of particular interest and obtain high-resolution images of them to add to my Leonardo slide deck!

But that was not the only reason why I eventually decided that, on my next cycling trip to Italy, I would make a point of seeking out that manuscript. While I was working on the Leonardo project, my focus was very much on Leonardo himself. The "book for merchants" manuscript Franci had discovered was simply the earliest copy-of-a-copy of many that happen to have survived to this day (or at least, survived and been found and identified as such).

After *The Man of Numbers* was published, however, my viewpoint began to shift away from Leonardo to a present-day retrospective view of the chain of events his work set in motion.

Reading William Goetzmann's work on the connection between *Liber abbaci* and the modern financial world, and then communicating with him by email, accelerated that shift. From my new perspective, that "earliest known abacus book" began to seem much more significant *as a cultural artifact* than I had previously believed. This was particularly so because I was beginning to consider using my project diary as the basis for a book about the process that led to *The Man of Numbers*. For that new project, it was clear to me that I would have to examine the missing-link manuscript as an important historical artifact.

For, after Franci's discovery, what was never in any doubt, to me or any other scholars, was the enormously significant role played in the arithmetic revolution by the shorter, simplified version of *Liber abbaci* that Leonardo tells us he wrote for merchants and businessmen. [As I noted earlier, in the 1228 edition of *Liber abbaci*, he calls it his *liber minoris guise*, or "book in a smaller manner." He also refers to it elsewhere.]

The problem for the historian was that no one knew exactly what was in that book. Based on the evidence that scholars had managed to assemble, it most likely comprised of material from the first ten chapters of *Liber abbaci*, together with parts of one of Leonardo's other books, *De Practica Geometriae*. But no one knew for sure. And without detailed knowledge of its contents, it was not possible to directly connect the myriad of abbacus books to anything Leonardo wrote.

Then, in 2003, Franci published her bombshell paper describing results of a study of a manuscript she had come across in the Biblioteca Riccardiana in Florence.

The manuscript is anonymous, Franci explains, and occupies pages 1 to 178 of the library's Codex 2404. The work itself is undated, but dates in some of the problems place its writing at around 1290, and the vernacular language used places it in the Umbria region. It is the only known pre-fourteenth-century abbacus book.

Known as *Livero de l'abbecho*—"Book of calculation"—from the anonymous author's introduction, the work is divided into 31 short chapters. It is definitely not an original volume. Roughly three-quarters of the problems are faithful translations into vernacular

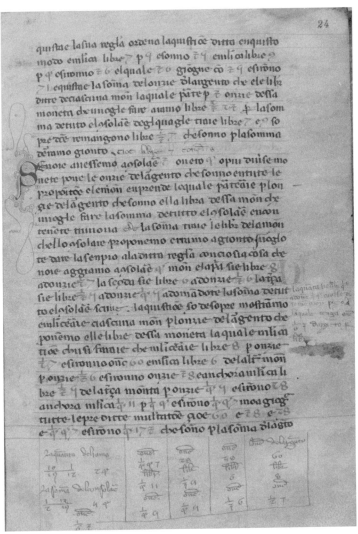

FIGURE 21. A page from the earliest known abbacus book, in the collection of the Riccardiana Library in Florence.

Italian of problems found in *Liber abbaci*. (Of particular note to a modern reader, it includes Fibonacci's famous rabbits problem, though recast in terms of pigeons.)

It was evidently written to be read by an audience (possibly just the writer himself) without significant training in mathematics, since the author began each problem type with simpler problems than those in *Liber abbaci*.

Could this be a copy of Leonardo's lost "book for merchants"? Or did the unknown author get the material directly from a copy of *Liber abbaci*? The former was always the more likely. Franci's analysis demonstrated that it had to be the former.

Since the writer gave no indication of having any particular mathematical skill, it seemed unlikely he could have carried out the difficult task of simplifying the sophisticated descriptions found in *Liber abbaci*. Moreover, his text included material on geometry that is hardly mentioned in *Liber abbaci*, though it could have come from Leonardo's *De Practica Geometriae*.

Further evidence in favor of the Umbrian's book being a derivative copy of the "book for merchants" was its inclusion of three chapters on calculating interest and depreciation, a topic not covered in *Liber abbaci* at all. This material was obviously included to make the treatise more useful to merchant readers, but where did it come from?

What we had, then, was a manuscript written around 1290, that provides a greatly simplified account of the new, and for the times highly sophisticated, mathematics in two of Leonardo's long, dense scholarly texts, coupled with some additional material on financial matters.

The mathematically unskilled author of the manuscript must surely have copied it directly from a single source. It is, in short, a medieval equivalent of a photocopy. But a photocopy of what?

There was only one mathematician at that time who had the ability to write the photocopy's source: Leonardo. Which means that the manuscript in the Riccardiana Library is not only the oldest known textbook on practical arithmetic in the Western world; it must in fact be a copy of Leonardo's lost "book for merchants." That was Franci's initial conclusion. It was subsequently confirmed by other scholars who studied a number of abbacus texts written in the first half of the fourteenth century. Although they make no meaningful references to Leonardo, they exhibit striking similarities with the Umbrian's treatise.[1]

All that information I obtained from Franci's article. Now, with *The Man of Numbers* behind me, I wanted to see the work for myself.

In the summer of 2011, I took yet another cycling vacation in Tuscany, staying in a small vacation rental apartment just outside Bunconvento, a few miles to the south of Siena, from which I could ride directly up to the mountain-top walled city of Montalcino, and from there on to other hilltop walled cities such as Pienza or Montepulciano.

On July 4, I reduced my daily cycle ride to a single hour early in the morning in order to drive up the Siena-Firenze Autostrade to visit the Riccardiana Library and see for myself the oldest known textbook on commercial arithmetic in the West, the content of which was almost certainly the text of Leonardo's lost "book for merchants."

A quick search on Google told me that the Biblioteca Riccardiana was located at number 10, Via Dè Ginori. Switching to Google Maps, I saw that it was about 300 meters north of Florence's famous Duomo.

As on a few other occasions when I had visited the center of Florence, I avoided having to drive through the ocean of photograph-taking, wandering tourists that flood the center, by driving as far as the city walls, seeking a parking space just outside one of the gates, and proceeding the rest of the way on foot.

---

[1]   I elaborate more on those studies at the end of Chapter 8 of *The Man of Numbers*.

Strolling along in shorts and a t-shirt (it was a very hot day), with a small bag over my shoulder, I was indistinguishable from all the other tourists. The only difference in my behavior was that I was not stopping to take photos of any of the many famous Florentine sights I passed. I had done the whole tourist-in-Florence thing many years earlier, during my first, lengthy stay at the University of Siena.

True, I had brought a small camera with me, but that was to take pictures of the exterior of the Riccardiana Library. (I knew I would not be permitted to take photographs inside the library.)

I had no difficulty finding the library. It was an entirely unremarkable building with an unadorned, pale yellow, stuccoed wall. There were two large, solid wooden doors surrounded by identical heavy stone arches, numbered 10 and 11. An engraved stone tablet proclaimed that this was indeed the Biblioteca Riccardiana, and beneath it a smaller engraved marble plaque gave the library's opening hours—a design feature that speaks volumes to the Italian assumption that most things do not change.

In front of door number 11, which was closed, was a sign saying that was the entrance to the section of the Biblioteca Riccardiana accessible to the public. I was sure that was not my door. So I stepped out of the hot, busy street through the adjacent open doorway, number 10. It was pleasantly cool inside, a result not of air conditioning but of centuries-old, thick, stone walls.

I immediately found myself in a rather dingy entrance hall, where a man sat in a small, glass-paned booth. As I expected the moment I saw him, he spoke no English. Unfortunately, my never-good Italian had gotten rusty since the days when I went to Italy more frequently, so I used one of a few rehearsed Italian sentences to explain to him that I had come to examine a particular manuscript.

Clearly, one of his main functions was to prevent tourists from entering the wrong door. "Niente visitare," he said. That much I understood: "No visitors!"

I repeated, in still slower Italian, that I wanted to see a particular manuscript.

Then I showed him the notebook I had brought with me, giving the reference number for the work I was interested in: Codex 2404, pages 1 to 178. Though this surely indicated that I was not a casual visitor looking to escape the early afternoon heat by gazing through glass on some ancient treasures, he repeated that visitors were not allowed.

Words such as "professore" and "matematico" did not work, nor did my prepared self-introduction, "Sono un professore di matematica da America"—in part, I suspect, because all visible clues classified me as one of the not-to-be-admitted tourists I had left behind in the outdoor heat.

The custodian seemed amicable enough, and I suspected his seeming curtness was more a matter of our having virtually no vocabulary in common—but my efforts continued to be of no avail.

At that moment, a fairly well dressed, scholarly looking man was just leaving the building. Witnessing my valiant attempts to explain my purpose, he came over and said, in English, "They do not allow visitors."

"Ma, io sono professore," I replied. The man at the desk repeated what he had said before, but again I did not understand. The professorial gentleman translated. "Do you have your university identity card?" Apparently the custodian had understood my "Sono un professore" routine, but I had been unable to understand his simple answer.

Having learned my lesson from my previous attempts to examine ancient manuscripts in Italy, on this occasion I had come well prepared. I took my wallet from the pocket of my shorts and pulled out my Stanford photo ID.

Presumably being totally unfamiliar with a Stanford University identity card, the custodian was now clearly trying to decide what to do with me. It could go either way.

"Momento," I said, rifling further through my wallet and this time extracting a white plastic card carrying the green logo of the Biblioteca Nazionale Centrale Firenze. This was the library card that had been secured for me back in 2004 by Jeannie Nesi (née Bettaney) from Manchester, the nanny turned librarian in the Rare Manuscripts section of the Florence National Central Library. In black lettering, the card gave my name and library membership number, together with the inscription "Sala Manoscritti," the verification that I had been granted access to the manuscript room of the BNCF.

I showed it to the custodian. "Questo buono?" I asked. From the look on his face, what I was showing him was clearly well beyond "buono." His face now beaming, he immediately motioned for me to proceed though the doorway leading into the library itself.

My BNCF card having got me through the entrance foyer—thank you, Jeannie—I climbed a rather dingy staircase and through the entrance to the collection room on the first (i.e., second) floor. It turned out there were still a couple more hurdles to overcome, but they were routine.

First, I had to put my small bag in a locker. The bag contained my annotated notebook to guide my examination of the manuscript, together with a street map of Florence, my spectacles, and a couple of pens. I was allowed to remove, and hold on to, my spectacles and my notebook, but not my pen. That would be a problem. How could I take notes?

Still, I duly deposited my bag, with its pens, an empty spectacle case, and my street map of Florence into a locker, and walked through to the collections room, with the voluminous pockets of my shorts containing an iPhone (with camera), a small Canon digital camera, my wallet, my passport, and the keys to my rental car—but sadly, no pens. Step one completed.

Being deprived of my pen proved to be no problem at all, however, for once inside the collections room, I was greeted by a curator who sat at a desk on which there was a large pot of pens, one of which she handed me to fill in the card to register me as a library user, and which I simply held on to after I was done. My Stanford ID, my passport, and my BNCF library card were duly inspected, details were filled in on various forms, and that was the second hurdle out of the way.

The main reading room of the library is spectacular, long and relatively narrow, with a high vaulted ceiling, decorated with a large fresco and a great deal of gilt molding. Books line both long walls, with walkways halfway up to permit access to the upper shelves. Above the top row of books on each long wall, where the ceiling starts to curve, are two relief coats of arms.

FIGURE 22. The reading room of the Riccardiana Library in Florence.

"Please, take a seat," the curator said when all the form-filling was completed and I had filled in a slip saying which manuscript I wanted to examine. She was pointing to one of the seats at the long manuscript reading table that ran down the center of the room. That could mean only one thing. I was about to lay my hands on the missing link.

Within a few minutes, a faded-beige-covered, bound volume was brought in from a room that was out of sight behind the archivists' desk, and it was placed on the table before me.

The volume was a little less than two inches thick. I knew from my earlier research that it contained two manuscripts, the second one not about mathematics.

With that, I opened the book and gazed at the first page of the oldest known "modern" arithmetic textbook in the Western world, written around 1290, and almost certainly a fairly faithful copy of Leonardo's own "book for merchants."

I had no trouble reading the vernacular Italian text. "This is the book of abacus according to the opinion of master Leonardo of the house of sons of Bonacie from Pisa," the unknown Umbrian author began.

In modern terminology, I was holding in my hands a "photocopy"— perhaps (though not likely) the first photocopy, but certainly the only one known today—of the book (Leonardo's "book for merchants") that made possible the world we all live in today. Because of this, we have our taken-for-granted supermarkets, bank accounts, credit cards, mortgages, car loans, technology, medicine, communications, transportation, entertainment technologies, international trading empires, and all the rest.

Acceptable academic terminology cannot capture the powerful emotional reaction I experienced. This was a moment like nothing I have ever experienced in my life, and perhaps never will again. For reasons I cannot fully comprehend, this particular manuscript had a far greater effect on me than had the two copies of *Liber abbaci* I had examined several years earlier.

Was the difference that it took me two attempts, a year apart, to see those? Was it that many people had examined the *Liber abbaci* manuscripts, and long before I saw them I had read their reports on what they found? Because, for all I knew, apart from Franci I was the first person to read these pages since the thirteenth century. Or was it that all of those small, vernacular abbacus books were the things that really fueled the arithmetic revolution? Whatever the explanation, my initial response surprised and overwhelmed me.

I spent about an hour, slowly leafing through the pages, not so much reading as experiencing the moment. (I already knew the book's contents, from reading Franci's account.)

Outside, thousands of tourists were flocking along the streets of Florence to see, and to photograph, what they believed were some of the greatest treasures of human civilization. And in that judgment they were correct. But there, inside the cool, quiet elegance of the manuscript reading room of the Biblioteca Riccardiana, I *held in my hands* a far greater treasure: the only known artifact in existence that tells us how all the rest came to be, and made possible much of what we know and experience today.

So what exactly did I find in the first half (actually, closer to two-thirds) of Codex 2404?

As I already knew from Franci's article, the manuscript is anonymous, and written in vernacular Italian. There is writing on both sides of the pages, so the book has 356 pages in modern numbering convention. Unlike the large pages of the *Liber abbaci* manuscripts I had seen in Siena and Florence, the pages of this work were the size of a modern book. The parchment was thin, some pages felt slightly brittle, and many pages were shiny. It was, however, in excellent condition, and the clear, rectangular lettering was easy to read. The author had ruled faint horizontal and vertical margins as well as lines on which to write the text.

The main text is brown, with all numerals in red. The section-opening capitals are either red with a blue background design or blue with a red background design. The first few words of a new section (often a problem statement) are in red.

It appeared that the book had been used as a textbook, with marginal annotations on many pages, in a smaller, very fine hand, either red or brown. It was not clear to me if the marginal notes were those of the text's author. To my archivally untrained modern eye, the handwriting looked similar. It certainly does not seem unreasonable to assume that the author began by copying Leonardo's text verbatim, without understanding (I so wanted it to be from Leonardo's original), and then worked through it to teach himself arithmetic, working out details in the margins.

Overall I spent about an hour going through the book, getting a sense of its contents, and trying to imagine myself in the thirteenth century setting out to learn elementary arithmetic and geometry from it.

Before leaving, I ordered copies of pages 1a, 24a, 121a, 140a, 142a, 145a, 153b, 155b, having decided those were most representative of the entire work.

And then I left—a very different person than the one who had entered just a bit more than an hour earlier.

See Florence and die, the saying goes. Well, I had just seen something that, for me, had an impact far exceeding anything to be found on the streets outside of the Reading Room of the Riccardiana Library.

CHAPTER 14

This Will Change the World

It's always hard, and perhaps even impossible, to know what it must have felt like to live through a major revolution in human thought or society that occurred many generations in the past. It is the nature of such revolutions that they change the very way we view life, making it impossible to imagine what it must have been like to grow up in a world that did not have whatever it is the revolution introduced.

The introduction of human language clearly was one such revolution. We literally cannot imagine a world without language.

Likewise, can we ever truly imagine a world where there were no numbers, as was the case prior to 8,000 years ago? I very much doubt it. Even the most mathematically averse and innumerate among us uses and depends upon numbers all the time. Try as I might, I could not put myself in the position of a thirteenth-century merchant—or even a thirteenth-century mathematician—reading *Liber abbaci.*

What we can do, on occasion, is draw a parallel between an earlier revolution with one that is taking place now, or has happened in the very recent past, within our memory. Of course, this is possible only if you are lucky enough to be alive during such a revolution. In my case, I was. In the 1980s, a revolution occurred that bears an uncanny similarity with Leonardo's introduction (to Western society) of the Hindu-Arabic number system and the methods of modern arithmetic. It was when I recognized this remarkable parallel that I got a sense of the impact *Liber abbaci* must have had.

Both revolutions involved computation, and both were in large part initiated by a single individual who, by having a keen eye for the marketplace (for ideas), played a catalytic role, rather than an inventive one. The second revolution I have in mind is the introduction of the personal computer, and in particular the Apple Macintosh in 1984. That revolution was chronicled by the technology writer Steven Levy in his 1994 book *Insanely Great: The Life and Times of Macintosh, the Computer that Changed Everything.*[1] I re-read Levy's book when I realized that the introduction of the Macintosh—and specifically the now familiar WIMPS interface (Windows-Icons-Mouse-Point-Select)—bore many similarities to the appearance of *Liber abbaci*. When I did so, I was astonished to discover just how strong the parallels are.[2]

---

[1]   Penguin Books, 1994.
[2]   The remainder of this chapter is adapted from the e-book companion I wrote to *The Man of Numbers*, titled *Leo and Steve: The Young Genius Who Beat Apple to Market by 800 Years*, an Amazon Kindle eBook original (2011).

PARALLEL LINES

PALO ALTO, CALIFORNIA, DECEMBER 1979

"Why aren't you doing anything with this? This is the greatest thing! This is revolutionary!"

The speaker was a young man aged 24 named Steven Jobs. Three years earlier, he and a friend, both computer hobbyists, had formed a fledgling computer company called Apple.

He and a group of his colleagues had just been given a tour of the famed Xerox Palo Alto Research Center, where PARC researchers had shown them a revolutionary new kind of computer called the Star. As part of a stock trade, Jobs had persuaded the PARC bosses to let him have a close look at this revolutionary new kind of computer people were whispering about.

Jobs had always felt that computers should be cheap, available to everyone, and above all easy to use. They were none of those. His company's own Apple II compßuter was selling for under $1500, less than a tenth of the cost of Star they had just seen

BUGIA, MAGHREB, N. AFRICA
ca. 1190

"Why aren't you doing anything with this? This is the greatest thing! This is revolutionary!"

The speaker was a young man aged 18 named Leonardo of Pisa. Two years earlier, after finishing school, he had sailed to Bugia to join his father, who had taken a post there as a customs agent.

Leonardo's father had invited him to Bugia in part to continue his education. During his time in north Africa, Leonardo had met and talked with the Arab scholars there, as his father had intended, and had learned from them the remarkable system for writing numbers and doing arithmetic that they had acquired from the Indian mathematicians who had developed it.

Growing up the son of a successful merchant in the major trading city of Pisa, Leonardo knew how important numbers were. But he also knew that it took many years of training to learn how to operate an abacus in order to do arithmetic. The schools did not

PALO ALTO cont.

in action, but you still had to be a dedicated computer nerd to use it. The only way to get the computer to do anything was to type in a string of barely comprehensible instructions at a keyboard.

The computer they had just seen was totally different. To anyone familiar with today's PCs, running Microsoft *Windows* or its precursor, the Macintosh Operating System, the Star would look very familiar. All of the basic components of today's PC interface were there—a mouse, a keyboard, and a display with icons and windows that you could open, close, move, and shuffle the contents with simple, mouse-controlled, pointing and clicking operations.

Jobs and his colleagues knew at once they had hit the mother lode. If Xerox doesn't see the potential in this thing, that's just too bad for them, he thought. From the moment the group left the lab, Jobs knew that the next computer Apple would bring out would be a cheap, affordable version of the Xerox Star.

Four years later, in 1983, the Macintosh was launched. Not long after that, Microsoft brought out its version of the Macintosh system,

BUGIA cont.

teach this skill. Leonardo had learned it by watching the accountants at work in the back rooms of the *fondacos*, while his father met with the other merchants in the front. He had practiced it himself many times, but still found it difficult.

But with the Hindu system, it was very different. Anyone today would recognize it as the system they learned in school. Writing and reading numbers was very simple. Just ten basic symbols and everything else was taken care of by the position they were in—units column, tens column, hundreds column, etc. Adding, subtracting, multiplying, and dividing were all done on a writing tablet by following simple rules. There was no need for an abacus.

Leonardo knew at once that this method for doing arithmetic could revolutionize trade and commerce. The European traders seemed unaware of this new system the scholars were studying. They continued to use an abacus to do calculations. Leonardo would change all that.

In 1202, after he had returned to Pisa, Leonardo released his book *Liber abbaci*, in which he described the Hindu number system and showed

PALO ALTO cont.

designed to run on IBM's personal computers. They called their system *Windows*. Suddenly, computers were no longer machines for math experts. With a bit of practice, anyone could use one.

The world would never be the same again. In an era where information was becoming increasingly important, the desktop PC, with its easy-to-use Windows interface, would give everyone access to it. Within a few years, the Internet and its successor, the World Wide Web, would change everything.

But why did it take a brash young computer fanatic like Jobs to make this happen? After all, the ideas for this new way to design computers had been around for some time. In 1968, before a packed audience at the Joint Computer Conference in San Francisco's Civic Center, Douglas Engelbart, a researcher at the Stanford Research Institute, had demonstrated an early version he had developed. And in 1973, PARC researchers had built a more sophisticated version called the Alto. Everyone who had seen these systems working thought they were very clever, but the new method never found its way out of the research labs and into the marketplace.

BUGIA cont.

how to use it. With this system, you did not have to deal with cumbersome Roman numerals or become skilled at using an abacus. With the Hindu system, after a bit of training, anyone could do arithmetic.

The world would never be the same again. In an era of trade and commerce where numbers were central, the Hindu system would give everyone access to them. Within two centuries, the scientific revolution, which would have been severely hampered without an efficient way to do arithmetic, would change everything.

But why was Leonardo's book the one that did this? After all, in the ninth century, al-Khwārizmī had written a book describing the system. Around 60 years earlier, Adelard of Bath had translated al-Khwārizmī's arithmetic book into Latin. So too had other scholars, among them Gerard of Cremona. While Leonardo was growing up, the Italian city of Toledo became an acknowledged center where scholars from all over the world came to study the works of the Arabs. Everyone who saw the Hindu system could see at once how much better it was than anything else that was available. And yet it never found its way out from the scholars and into the hands of the merchants.

PALO ALTO cont.

Apple programmer Andy Hertzfield was one of the group who visited the PARC lab on that fateful day in 1979. Speaking about the visit some years later, not long after the Macintosh came out, he said:[1] "There are two barriers that keep 150 million people from using the computer. First, it's too expensive.

Second, it's too hard to use. You have to get immersed in this muck of horrible stuff. Computers are great, but they don't do any good if it doesn't reach the common man. But we're bringing computers to the people for the first time. When we designed Macintosh, we aimed at ourselves—people. . . . We want the man on the street to get a Mac and feel that incredible potential . . . it's going to change the world."

Writing about these events in 1994, author Steven Levy explained it this way:[2] "By and large, they [the PARC researchers] viewed themselves as pure researchers. A prototype, a paper, an article in *Scientific American* . . . those were their products."

BUGIA cont.

The answer in part is that the merchants did not see a pressing need for a new system. There was no shortage of skilled abacus users to do the calculations for them. It probably never occurred to them that it would be much more efficient if they could do the calculations themselves.

To do arithmetic meant having to become immersed in Roman notation for writing numbers and learning how to operate an abacus. Yes, numbers were great, but handling them was viewed as a highly specialized activity often left to the experts. Leonardo saw things differently. He saw that, with the Hindu system, numbers and arithmetic could be put in everyone's hands. And doing so would change the world.

The European scholars did not see it that way. They were not merchants or traders. They dealt with ideas. For them, the Hindu number system was an interesting object of study, something to discuss and to write about—for each other.

[1]   Quoted in Levy, 1994, pp. 23–24.
[2]   Ibid., p. 72.

PALO ALTO cont.

PARC researcher Butler Lampton expressed it in these words:[3] "The purpose of PARC was to learn.... But it wasn't critical that Xerox develop those ideas ... the main product of a research laboratory is ideas."

The young men from Apple Computer, however, saw things very differently. Here is how Levy summarizes their attitude:[4] "[Their] job was to create new products that would both change the world and bring in some cash. The ethic was more practical: If it didn't hit the streets, it wasn't worth doing. Ideas were useless if they didn't get out there. For the Macintosh, the most important design consideration was getting it on people's desks.

As for new ideas, yes, the Apple folks cooked up a thick stew. But the best ones were borrowed. These were, of course, the ideas concocted and nurtured by the computer science Illuminati at the Xerox Palo Alto Research Center."

BUGIA cont.

They had all read al-Khwārizmī, either in the original or in translation. Hindu arithmetic was a marvelous piece of mathematics, everyone agreed. An admirable piece of scholarship.

Leonardo saw things differently. His book would not be written for the scholars. He was going to target it to the traders and merchants—people like his father and the other merchants he had observed while growing up in Pisa. That would mean writing it in a very different fashion from the standard scholarly texts. In addition to describing the system, he would have to present lots of examples showing how to use it to solve real-life problems.

Leonardo was an excellent mathematician, and his book would contain many new ideas. But the main value of *Liber abbaci* would be that it brought into the marketplace the ideas developed by the brilliantly creative Hindu mathematicians.

---

[2]   Ibid., p. 73.
[3]   Ibid., p. 74.

PARALLEL LINES MAY NEVER MEET, BUT THEY LOOK ALIKE

Of course, you can take any analogy only so far. Parallel lines never meet. There are a number of differences between Leonardo's publication of *Liber abbaci* and the manufacture of the Apple Macintosh computer (more generally, the introduction of a completely new way of operating computers). For one thing, the timescales were very different. In the thirteenth century, things happened slowly, over many decades; in the late twentieth century, hardly anyone could keep up with the pace of change.

Moreover, I am certainly not saying that Steve Jobs was the Fibonacci of the twentieth century. The analogy is between the two events and the changes they led to, not the individuals involved. Leonardo wrote *Liber abbaci* on his own. In contrast, although Steven Jobs's initial drive to develop the Mac was crucial to making it happen, in the end a great many individuals were involved in its design and manufacture.

But these are just details (unless you were an early Apple employee, that is). Parallel lines may never meet, but they certainly look alike—they are, after all, *parallel*. As I gradually pieced together a picture of Leonardo and the events that led up to his publication of *Liber abbaci*, I was struck time and again by the strong similarity between those two revolutions, 800 years apart. When you allow for the very different timescales over which change occurred, many more parallels emerge.

For instance, when *Liber abbaci* came out, several decades elapsed before there was any real competition (as far as we know). In contrast, as soon as the Macintosh hit the retail stores, many companies introduced copies. One of those copies would soon eclipse the Mac and dominate the PC market. Just over a decade later, 95% of all personal computers in the world would run an operating system called Windows, a Macintosh rip-off produced by a company called Microsoft, founded and run by a fiercely ambitious young software entrepreneur named Bill Gates. The Macintosh would be marginalized to a few specialist corners of the computer market. (The first version of Microsoft Windows came out in 1985. Windows really took off in the early 1990s with the appearance of Version 3.1. The Macintosh finally achieved a significant market share after Jobs returned to run Apple in 1997.)

But when you allow for the different timescales, the parallel holds up very well. Inspired by Leonardo's work, over the succeeding centuries hundreds of other authors produced derivative works, ranging from slavish translations of parts of *Liber abbaci* into vernacular Italian to works that shared little that was specifically in common with Leonardo's text. These were the abbacus books, and they came to dominate the market, so that by the fourteenth century it appears that hardly anyone ever consulted the book that started it all. So, yes, the abbacus books were the Windows machines of the medieval period.

For those of us living today, I think this analogy can help give a sense of what it would have been like to have stood by Leonardo's shoulder as he was working on his book. In fact, the more you think about it, the more you realize the parallels are even stronger than I have indicated so far.

First, both advances were essentially *interface* issues. The Hindu mathematicians did not invent numbers; they developed a new way of writing them down and working with them. Likewise, the pioneers of WIMPS computing did not invent computers; they developed a new way of operating them and getting information from them.

Both developments were also about providing more efficient interfaces to *abstractions*: numbers are abstractions and so too are the bits and bytes that constitute the information stored in a computer.

By providing efficient interfaces, both the Hindu number system and WIMPS computers handed control to the user. The Italian merchant no longer had to give the figures to his accountant and wait until the answers were available; he could do it for himself, on the spot. Similarly, the person who sat down at a WIMPS computer no longer had to type in some obscure command sequence and wait there while the computer carried out its processes and then returned the answer. The user was given control of the information, and could manipulate it directly, using the mouse and the keyboard.

Moreover—and I make this point in case you are wondering if this entire chapter is not making altogether too much fuss about something mundane and ordinary—both advances resulted in systems that were so natural and easy to use, people soon forgot what a huge step each had been, and how much each changed life on Earth. Today it is hard to imagine a world without numbers or WIMPS computing. But such worlds did once exist, one prior to 1200 CE, the other within my own living memory!

Before the introduction of Hindu numerals, arithmetic was difficult. The system for writing numerals was cumbersome, and computation required mastery of complex finger reckoning or the abacus. Likewise, prior to the introduction of WIMPS computing, using computers was difficult. Getting information in and out was cumbersome. You had to learn arcane computer languages in order to make them do anything. Computers were for experts—often self-taught, and generally very young.

With Hindu arithmetic—which many Europeans (Leonardo was not one of them) soon began to refer to as Hindu-Arabic arithmetic—anyone could handle numbers. The efficiency of the new system helped make possible the scientific and technological revolutions that followed. With WIMPS computing, anyone could operate a computer. The simplicity and efficiency of the new system made possible desktop computing (in the office and the home), word processing, desktop publishing, email, the Internet, the World Wide Web, and Web commerce.

For both Leonardo's introduction of Hindu-Arabic arithmetic and the introduction of WIMPS computing, there were two key steps: first, the recognition that the new idea had great practical application; and second, the ability to take the new idea and package it in such a way that the right people had access to it and could use it. In both cases—numbers and computers—many others had seen the same system but failed to recognize its significance. (Or they failed to do anything about it, which amounts to the same thing in terms of changing the world.) In the end, it was Leonardo who completed the two steps for Hindu-Arabic arithmetic and a group of young computer fanatics, iconicized and to some extent led by Steven Jobs, who did it for WIMPS computing. Neither Leonardo nor Jobs (and later the abbacus authors and Gates and others) came up with the key ideas themselves. Their role was seeing the potential and following up with improvement, packaging, and the right marketing.

CHAPTER 15

Leonardo and the Birth of Modern Finance

While I was putting the finishing touches to *The Man of Numbers*, I came across an article about Leonardo that presented a view of his work that I had not been aware of: his role in the birth of modern finance. The author of the article, Professor William N. Goetzmann of Yale University, had, like me, read Sigler's English translation of *Liber abbaci* soon after it appeared. Whereas I read Leonardo's text through a mathematician's eyes, Goetzmann, who directs the International Center for Finance in the Yale School of Management, brought to it the perspective of a world expert in finance. His article shows that the financial system we take for granted in today's world is yet another echo of Leonardo Pisano.

Goetzmann wrote: "The five-hundred-year period following the appearance of *Liber abbaci* saw the development in Europe of virtually all the tools of financial capitalism that we know today: share ownership of limited-liability corporations, long-term government and corporate loans, liquid and active international financial markets, life insurance, life-annuities, mutual funds, derivative securities, and deposit banking. Many of these

FIGURE 23. William N. Goetzmann, Edwin J. Beinecke Professor of Finance and Management Studies and Director of the International Center for Finance, Yale University. Photograph courtesy of William Goetzmann.

developments have their roots in contracts that were based on the mathematical analyses Leonardo introduced to Western Europe through *Liber abbaci*."[1]

According to Goetzmann, evidence in *Liber abbaci* (1202) suggests that, in particular, Leonardo was the first to develop present-value analysis for comparing the economic value of alternative contractual cash flows.

Present-value analysis is a method for comparing the relative economic value of differing payment streams, taking into account the changing value of money over time. Mathematically reducing all cash flow streams to a single point in time allows the investor to decide which is unambiguously the best. According to a 2001

---

[1]   Goetzmann, 2005, p. 125.

survey of corporate financial officers,[2] the present-value criterion is now used by virtually all large companies in capital budgeting decision making. The modern present-value formula was developed by economist Irving Fisher in 1930,[3] but, according to Goetzmann, its origins can be found in *Liber abbaci*.

When I read Goetzmann's article, I knew I would have to include an account of his analysis in this second book. But there was a problem. Financial mathematics is a highly specialized area in which, like many mathematicians, I have little experience. In contrast, Goetzmann is a distinguished expert in such matters. So I contacted him and asked for his assistance. With his kind permission, this chapter follows his article fairly closely.

Interestingly, both Goetzmann and I made the same observation, that *Liber abbaci* had such an enormous influence on the world, influencing the evolution of capitalist enterprise and public finance in the centuries that followed, because it met a particular commercial need at a crucial time in European history.

For instance, the complexity of new European commercial practices in the twelfth century required arithmetical tools for solving problems of conversion and exchange. A merchant in Pisa who traded goods just within Northern Italy, for example, had to be able to calculate currency conversions from the money of each city-state, a task that involved mastering the comparative values and differential silver content of Bolognese, Pisan, Venetian, and Genoese lira, and knowing their value relative to Byzantine besants, Imperial pounds, Barcelona lira, and Magalonese soldi.

Moreover, the need for new mathematics was by no means restricted to monetary exchange. The very factors that impeded European trade—from bad roads and a multiplicity of currencies, to varying legal standards and lack of credit markets—also created

2   Graham and Campbell, 2001.
3   Fisher, 1930.

opportunities for entrepreneurs who could surmount them. *Liber abbaci* provided those entrepreneurs with a wide range of practical mathematical tools they could use: methods for calculating present value, compounding interest, evaluating geometric series, dividing profits from business ventures, pricing goods and monies involving a complex variety of weights, measures, and currencies, and so forth.

On a more general level, Goetzmann says that *Liber abbaci* provided what amounted to a new mathematical approach to financial decision making, an innovation that, in turn, allowed European mathematicians to value increasingly complex financial instruments in the centuries that followed.[4]

Leonardo's treatment of financial mathematics built on a long history of texts explaining how to carry out financial calculations, but had important novel features not found in earlier works.

At least seven centuries before the Pisan, Indian mathematicians were calculating interest rates and investment growth. For example, one of India's earliest mathematicians, Ryabhata (475–550), presented and solved some interest rate problems in his famous book *Ryabhatīya*, a work better known for its contribution to astronomy. In the seventh century, Bhāskara (ca. 600–680) wrote an extension and commentary on *Ryabhatīya* that contained several practical applications of the mathematics in the earlier work, including partnership share divisions, and the relative pricing of commodities. And Sridharacarya's (ca. 870–930) tenth-century *Trisastika*, a work of 300 verse couplets, contained some very practical interest rate problems and a division of partnership problem.

Closer to Leonardo's time, and similar in spirit to the financial problems in the Pisan's work, is the *Lilivati* of Bhāskarācārya

---

[4] Goetzmann, 2005, pp. 123–25. Goetzmann actually suggests that Leonardo *invented* the approach, but this seems hard to sustain. Given the commercial environment in which Leonardo wrote *Liber abbaci*, however, it is highly likely that his treatment was novel, and included techniques that had not been described before.

(1114–1185), written around 1150. Like the earlier works *Trisastika* and *Ryabhatīya*, *Lilivati* contained loan problems and methods of finding principal and interest.

There is no evidence that Leonardo consulted any of those Indian texts, but by way of Arabic manuscripts he was likely aware of the ideas they presented. In particular, many of the methods and problems in *Liber abbaci* came from al-Khwārizmī's *al-jabr*. The principal difference between Arabic texts and the Italian ones lay not in the methods described, but in the problems. Arabic works tended to focus on legal issues related to legacies and dowries, such as the division of assets among family members and other claimants. Italian algebra texts don't contain such problems, because Christian legacies did not pose the mathematical challenge that Islamic legacies did. The Qu'ran stipulates rules governing the division of an estate, and the calculations can be complicated. On the other hand, certain problem types seem to be unique to Italian algebra works. In particular, usury is against the Islamic religion, so problems on interest are not found in the Arabic texts.

The following is a typical problem scenario in *al-jabra*:[5]

> A man dies, leaving two sons behind him, and bequeathing one-third of his capital to a stranger. He leaves ten dirhems of property and a claim of ten dirhems upon one of the sons.

Leonardo took the mathematical methods from *al-jabr* and adapted them to problems of dividing the capital from commercial ventures among partners, rather than among family members. Thus, the mathematics is essentially the same in the two works, but they differ in applications.

Goetzmann's article allows a twenty-first-century reader to view *Liber abbaci* from the perspective of modern financial mathematics.

---

[5]   Rosen, 1831, p. 86. Al-Khwārizmī does not actually pose a question, but rather proceeds directly to calculate the dead man's worth.

Liber abbaci Seen through the Lens of Modern Finance

Chapter 8 of *Liber abbaci* is entitled *On Finding the Value of Merchandise by the Principal Method*. The Principle Method was more generally known as the Rule of Three. The reader meets it in the first problem of the chapter, which asks for the price of a given quantity of merchandise when the price per unit is known: If 100 rolls costs 40 lira, how much would 5 rolls cost? Leonardo solves it by laying out the problem information in a diagram:

$$40 \quad 100$$
$$? \quad 5$$

from which he obtains the answer $(40 \times 5)/100$, i.e., 2 lira.

Leonardo used this Rule of Three with increasingly complex quantities and currencies, applying it to examples drawn from trade around the Mediterranean. Goods included hundredweights of hides, hundredpounds of pepper, tons of Pisan cheese, rolls of saffron, nutmeg, and cinnamon, meters of oil, sestario of corn, canes of cloth, oil from Constantinople. Currencies included denari, massamutini, bezants, tareni. Transactions were described that took place in Sicily, the Barbary Coast, Syria, Alexandria, Florence, Genoa, Messina, and Barcelona.

The selection of problems in *Liber abbaci*, and their presentation, were clearly designed so his book would appeal to merchants. For example, for a Pisan cloth merchant who traded in Syrian damask, and competed with a Genoese merchant, it would have been useful to be able to translate across three different units of length, and Leonardo duly obliges with this problem:[6]

A Pisan cane is 10 palms or four arms however a Genoese cane is said to be 9 palms. And furthermore the canes of

---

[6]   Sigler, 2002, p. 168.

Provence and Sicily and Syria and Constantinople are the same measure.

Another problem would surely get the attention of traders of raw cotton:[7]

> One has near Sicily a certain ship laden with 11 hundred-weights and 47 rolls of cotton, and one wishes to convert them to packs; because $^1/_31$ hundredweights of cotton ... is one pack, then four hundredweight of cotton are 3 packs and for rolls of cotton are 3 rolls of a pack; you write down in the problem the 11 hundredweights and 47 rolls, that is 1147 rolls below the 4 rolls of cotton and you will multiply the 1147 by three and you divide by the 4; the quotient will be ¼860 rolls of a pack.

Leonardo went on to examine exchange between two goods. For example:[8]

> It is proposed that 7 rolls of pepper are worth 4 bezants and 9 pounds of saffron are worth 11 bezants, and it is sought how much saffron will be had for 23 rolls of pepper.

For this problem, Leonardo drew a diagram with three columns:

$$\left(\begin{array}{ccc} saffron & bezants & pepper \\ \left[\dfrac{2\ 8}{7\ 11}10\right] & 4 & 7 \\ 9 & 11 & 23 \end{array} \right)$$

The calculation requires multiplying 23 x 4 x 9 and dividing by the diagonal elements 7 x 11 to arrive at the answer. As Goetz-mann observes, Leonardo's presentation makes it clear that "this

[7]   Ibid., p. 176.
[8]   Ibid., p. 184.

pattern can be applied to any longer sequence of intermediate trades to establish no-arbitrage relationships among commodities in a market."[9]

Although the Rule of Three seems (and is) trivial today, it was unquestionably a vital quantitative tool for thirteenth-century traders. According to Goetzmann:

> Italian merchants were buying saffron and pepper from their Arab counterparties, who were the primary intermediaries in the spice trade. Merchants who could not calculate the relative value of saffron and pepper in the market—or perhaps could only do so approximately, or with some difficulty— were at an extreme disadvantage in trade and negotiation. Just as today's hedge fund managers use sophisticated, quantitative models to calculate, and dynamically adjust, the relative price of mortgage-backed securities, so must thirteenth-century Pisan merchants, bartering in a Damascus *suk* with spice traders, have used arithmetic calculations to barter their way towards profit. It is thus not surprising that the knowledge of the Rule of Three extended in antiquity along trade routes.[10]

Leonardo also applied the Rule of Three to problems of currency exchange.[11] The practical demand for a guide to monetary conversions must have been considerable at the time. Italy had the highest concentration of different currencies in Europe, with 28 different cities at one time or another issuing coins, 7 in Tuscany alone. Most, but not all, of them were based on the Roman system of "*d,s,l*" denari, soldi, lire—familiar to older English readers as the pounds, shilling, and pence system. However, their relative value

[9]   Goetzmann, 2004, p. 17. (Also Goetzmann, 2005, pp. 131–32.)
[10] Ibid.
[11] Goetzmann, 2004, pp. 18–20. (Also in a slightly different form, in Goetzmann, 2005, pp. 132–33.)

and metallic composition varied considerably through time and across space. The varieties of currency created business for money changers, *Banche del Giro*, where monetary systems came into contact, and these money changers and their customers needed tools for calculation.

One complicating factor was that transactions might be expressed in a certain currency, or unit of account, but the quantity of physical coinage corresponding to that unit could vary through time, as governments regularly debased and re-valued their currencies. This made pure gold coins important, like the Florentine florin first minted in 1252, which could serve simultaneously as a unit of account and of transaction, with a 1:1 numeric correspondence. Most silver coinage of the time, in contrast, had at least some level of debasement with copper that caused its value relative to bullion to fluctuate through time. This explains why Leonardo's monetary analysis goes beyond currency conversions into problems of minting and alloying of money. (An entire chapter of *Liber abbaci* is devoted to the methods for minting coins of silver and copper.)

Certain cities—including Pisa—had imperial rights to coin money, and mints in these cities operated by striking coins from bullion brought to them from the public and the private sector. A merchant who wished to turn a quantity of metal (perhaps even old coins) into currency would pay a seigniorage fee to the mint for the production of coin.

During much of Leonardo's lifetime, the Pisan penny was pegged to the value of the Lucchese penny and together they became the standard monetary unit in Tuscany, with little evidence of debasement or changes in the metallic composition. In contrast, also during Leonardo's lifetime, the currency of Northern Italy underwent a revolution—cities began to mint large silver coins called *grossi*. In the late 1220s (around the time of the second edition of *Liber abbaci*), Lucca introduced a coin with a high silver content worth 12 denari. This followed the introduction of *grossi* by Genoa in 1172 and by Venice in 1192. Leonardo did not specifically address the relative value of *grossi* and *denari* on his chapter on minting of coin. Instead, he abstracted away from particular currencies to the general methods of calculation and evaluation of relative proportions of silver and copper in coins. Presumably the methodology developed in *Liber abbaci* was useful not only to the master of the Pisa mint, but perhaps to merchants who brought metal to be coined, and paid the seignorage.

GOETZMANN'S THESIS

Leonardo devoted a great deal of Chapter 12 of *Liber abbaci* to specific problems of finance. So rich is this material and its impact on the subsequent development of financial practice that, even with Goetzmann's expertise as my guide, as hitherto, I find myself unable to do the work the justice it deserves. As a mathematician, I can read Leonardo's words and understand his mathematics. What I cannot do, however, is see the world through the eyes of someone knowledgeable about the application of mathematics in trade, commerce, and finance. I was pleased, then, when Professor Goetzmann kindly agreed to allow me to present his own words for the remainder of this account. The rest of this chapter comprises pages 20 to 30 of the 2004 article by Goetzmann cited earlier, exactly as he wrote them (save the spelling of *abbaci*).[12]

Most of the pure finance problems in *Liber abbaci* are in the twelfth chapter, and fall into four general types. The first type concerns how the profits from joint business ventures should be fairly split when contributions are unequal and are made at different points in time, and in different currencies or goods

---

[12]   The same also appears, in a slightly revised form, in Goetzmann 2005, pp. 133–39.

and in cases in which business partners borrow from each other. The second type concerns the calculation of the profits from a sequence of business trips in which profit and expense or withdrawal of capital occurs at each stop. The third is the calculation of future values from investments made with banking houses. The fourth is present-value analysis as understood today, including specifically the difference between annual and quarterly compound interest.

Division of Profits

The mathematics of dividing profits from business joint ventures was obviously relevant to Italian merchants of the thirteenth century. The basic business unit used to finance many of the trade ventures in Northern Italy in the Middle Ages was the *commenda* contract between an investor and his traveling partner—the former (*commendator*) investing capital and the latter (*tractator*) investing labor. John H. Pryor's (1977) study of *commenda* from the twelfth century describes two basic types—a unilateral and a bi-lateral contract—the former providing limited liability but smaller profit to the *tractator* and the latter sharing the potential losses equally between the contracting parties.[13] In the standard, unilateral commenda, the *commendator* would transfer capital to the *tractator* for the duration of the voyage (or term of the contract) and would take 3/4 of the profit. Much of what is known about *commenda* has been gleaned from studies of notarial documents in Italian and French archives, and Pisan records are among the oldest. Pisa's *Constitutum Usus* (1156) is the earliest surviving municipal document specifying the conditions of the *commenda* contract.[14] The unilateral *commeda* in Pisa was very much like the

---

[13]   Pryor, 1977.
[14]   Sigler, 2002, p. 10.

contract described by Leonardo in a problem of division of profits, writing 44 years after the statutes. The problem *On Companies*, is worth quoting at length:[15]

> Whenever . . . any profit of an association is divided among its members we must show how the same must be done according to the above-written method of negotiation. . . . We then propose this of a certain company which has in its association 152 pounds, for which the profit is 56 pounds, and is sought how much of the same profit each of its members must be paid in pounds. First, indeed, according to Pisan custom, we must put aside one fourth of the above-written profit [apparently for the *tractator*]; after this is dealt with, there remain 42 pounds. Truly if you wish to find this according to the popular method, you will find the rule of 152, that is $\frac{10}{819}$; you divide the profit, namely the 42 pounds; by 8; the quotient will be 5 pounds and 5 soldi which are 105 soldi, which you divide by the 19; the quotient will be 5 soldi and $\frac{6}{19}$6 denari.

$$
\left(
\begin{array}{cc}
profit & capital \\
42 & 152 \\
\dfrac{6\ \ 6\ \ 5}{19\ 12\ 20} & 1
\end{array}
\right)
$$

> If you wish to find by the aforewritten rule how much will result . . . for profit from 13 pounds in the association, then you do this: multiply the 13 by the profit portion in one pound, namely by 5 soldi and $\frac{6}{19}$6 denari . . . there will be 3 pounds and 11 soldi and $\frac{2}{19}$10 denari.

15   Ibid., p. 172.

$$\begin{pmatrix} profit & capital \\ 42 & 152 \\ \dfrac{2\ 10\ 11}{19\ 12\ 20}\text{—}3 & 13 \end{pmatrix}$$

This problem provides a fascinating example of how *commenda* were used. Although the *tractator* profits were divided according to Pisan custom, the *commendator* shares could be divided among several investors, whose association is termed a *societas*. Fibonacci thus develops a general approach to dividing the profits from a *societas* according to the proportion of contributed capital. His approach is the now familiar definition of a rate of return on a unit of capital which is then scaled by the amount invested. Robert Lopez argues that *societas* and the *commenda* contracts were the original European forms of business association from which modern partnership and corporations ultimately developed. Consequently, the *commenda* example, and related problems addressed by Fibonacci in *Liber Abaci*, were significant early contributions to "pre-corporate" financial economics.

Traveling Merchant Problems

The second type of financial problem is a set of "traveling merchant" examples, akin to accounting calculations for profits obtained in a series of trips to trading cities. The first example is:[16]

> A certain man proceeded to Lucca on business to make a profit doubled his money, and he spent there 12 denari. He then left and went through Florence; he there doubled his money, and spent 12 denari. Then he returned to Pisa, doubled his money and it is proposed that he had nothing left. It is sought how much he had at the beginning.

---

[16]   Ibid., p. 372.

Leonardo proposes an ingenious solution method. Since capital doubles at each stop, the discount factor for the third cash flow (in Pisa) is 1/2 1/2 1/2. He multiplies the periodic cash flow of 12 denari times a discount factor that is the sum of the individual discount factors for each trip, i.e.,

$$\frac{1}{2} + \frac{1}{4} + \frac{1}{8}$$

The solution is 10½ denari. The discount factor effectively reduces the individual cash flows back to the point before the man reached Lucca. Notice that this approach can be generalized to allow for different cash flows at different stages of the trip, a longer sequence of trips, different rates of return at each stop, or a terminal cash flow.

In the 20 examples that follow the Lucca-Florence-Pisa problem, Leonardo presents and solves increasingly complex versions with various unknown elements. For example, one version of the problem specifies the beginning value and requires that the number of trips to be found—e.g., "A certain man had 13 bezants, and with it made trips, I know not how many, and in each trip he made double and he spent 14 bezants. It is sought how many were his trips."[17] This and other problems demonstrate the versatility of his discounting method. They also provide a framework for the explicit introduction of the dimension of time, and the foundation for what we now consider finance.

Interest Rate and Banking Problems

Immediately following the trip problems, Fibonacci poses and solves a series of banking problems. Each of these follows the pattern established by the trips example—the capital increases

---

[17] Ibid., p. 383.

by some percentage at each stage, and some amount is deducted. For example:[18]

> A man placed 100 pounds at a certain [banking] house for 4 denari per pound per month interest and he took back each year a payment of 30 pounds. One must compute in each year the 30 pounds reduction of capital and the profit on the said 30 pounds. It is sought how many years, months, days and hours he will hold money in the house.

Fibonacci explains that the solution is found by using the same techniques developed in the trips section. Intervals of time replace the sequence of towns visited and thus a time series of returns and cash drawdowns can be evaluated. Once the method of trips has been mastered, then it is straightforward to construct a multi-period discount factor and apply it to the periodic payment of 30 pounds—although in this problem the trick is to determine the number of time periods used to construct the factor. Now we might use logarithms to address the problem of the nth root for an unknown n, but Fibonacci lived long before the invention of logarithms. Instead, he solves it by brute force over the space of three pages, working forward from one period to two periods, etc., until he finds the answer of 6 years, 8 days, and $\frac{1}{2}\frac{3}{9}5$ hours. The level of sophistication represented by this problem alone is unmatched in the history of financial analysis. Although the mathematics of interest rates had a 3,000-year history before Fibonacci, his remarkable exposition and development of multi-period discounting is a quantum leap above his predecessors.

Eleven banking house problems follow this initial mathematical tour de force. Each of them has some value of the problem that is unknown—either the future value, the present value, the rate of interest, the duration of the loan, or the periodic dividend that is

---

[18]   Ibid., p. 384.

withdrawn. Fibonacci provides model solutions to each unknown, basing all of them on the previously introduced "trips" paradigm.

It is a bit surprising to find such an overt description of the charging of interest by bankers in the thirteenth century, given the ecclesiastical proscription against usury. John Munro has argued convincingly that, while Church doctrine had long forbidden usury, vigorous institutional attacks on the charging of interest began in earnest in the early thirteenth century and were a major impetus for the development of *census* and *rentes* and perpetual bond contracts—all of which circumvented the ecclesiastical definition of loans.[19] *Liber abbaci* was written just before the formation of the Franciscan and Dominican orders (1206 and 1216 respectively) who were at the vanguard of the fight against usury. The Church's attack on the charging of interest was in fact an attack on just the operations described and analyzed in Leonardo's book.

There is some evidence in *Liber abbaci* that Fibonacci's financial analysis may have roots in the ancient Near East. Consider this problem:[20]

> A certain man gave one denaro at interest so that in five years he must receive double the denari, and in another five he must have double two of the denari and thus forever from 5 to 5 years the capital and interest are doubled. It is sought how many denari from this one denaro he must have in 100 years.

Compare this to the problem from the Old Babylonian tablet 8528 in the Berlin museum published by Otto Neugebauer in 1935. It asks:

> If I lent one mina of silver at the rate of 12 shekels (a shekel is equal to 1/60 of a mina) per year, and I received in repayment,

[19] Munro, 2003.
[20] Sigler, 2002, p. 437.

one talent (60 minas) and 4 minas. How long did the money accumulate?

The solution is found by capitalizing interest only when the outstanding principal doubles. At 20%, the principal doubles every five years. Thus, the answer is that it would take 30 years for the debt to grow to 64 minae. The doubling over five-year intervals seems a striking parallel, particularly given the fact that an annual compound rate that yields that result is not a round number—it is something less than 15%.

Present-Value Analysis

The most sophisticated of Fibonacci's interest rate problems is *On a Soldier Receiving Three Hundred Bezants for his Fief*. In it, a soldier is granted an annuity by the king of 300 bezants per year, paid in quarterly installments of 75 bezants. The king alters the payment schedule to an annual year-end payment of 300. The soldier is able to earn 2 bezants on one hundred per month (over each quarter) on his investment. How much is his effective compensation after the terms of the annuity have changed? To solve this, Fibonacci explains:[21]

> First indeed you strive to reduce this problem to the method of trips and it is reduced thus . . . because there are 4 payments, 4 trips are similarly carried, and because each payment is 75 bezants, this is had for the expense of each trip. Next because 53 is made from 50, you put $\frac{50}{53}$ four times for the four payments thus
>
> $$\frac{50 \quad 50 \quad 50 \quad 50}{53 \quad 53 \quad 53 \quad 53}...$$

---

[21]   Sigler, 2002, p. 392.

The product of these ratios is used to discount a payment by four periods (or trips). Notice that Fibonacci's novel fraction expression allows each term to be successively discounted by one more period. The discounted annual value of the 300 bezants paid in the last period is 259 and change. As before, Fibonacci explains how to construct a multi-period discount factor from the product of the reciprocals of the periodic growth rate of an investment, using the model developed from mercantile trips in which a percentage profit is realized at each city. In this problem, he explicitly quantifies the difference in the value of two contracts due to the timing of the cash flows alone. As such, this particular example marks the discovery of one of the most important tools in the mathematics of finance—an analysis explicitly ranking different cash flow streams based upon their present value.

Institutional and Commercial Context

The soldier's present-value problem above testifies to the fact that the mathematics of time-value discounting was apparently important in 1202 when *Liber abbaci* was written. However, over the next few centuries its significance to governmental and commercial transactions grew tremendously. With the growth of international trade—and warfare—in the late Middle Ages, the need for long-distance monetary transfers increased. Chinese merchants had solved this problem centuries earlier with the use of *feichan* "flying money" remittance certificates issued by government institutions in provincial capitals. Arab merchants in Fibonacci's time used an instrument much like a modern personal check for paper remittance. Neither of these explicitly involved a discount for the time-value of money. Early European money transfers—bills of exchange—did. Any distant remittance involved a time interval due to travel, and compensation for the use of the money over that period is natural, so bills of exchange were typically discounted to compensate for forgone interest revenue—essentially as in the example of *The Soldier Receiving Three Hundred Bezants for his Fief.* Thus, bills of exchange were also interest rate instruments. Ashtor (1983) published the earliest known bill of exchange, which appears to date from 1220.[22] Sivéry identifies an Italian discount contract with an implicit interest rate of around 11% from the year 1252.[23] Both the early thirteenth century date and the Levantine origin of medieval bills of exchange form an interesting parallel to the publication date and Eastern influences of *Liber abbaci.* Were the mathematical tools for present value imported to Italy along with bills of exchange and Arabic numerals? Or did the technological innovation of using bills of exchange as debt securities (called "dry-exchange" by medieval practitioners) stimulate mathematical work on the time-value of money? In all likelihood, it was a fertile interplay between commerce and mathematical advancement, and Leonardo of Pisa was at its geographical and intellectual center.

Another key innovation occurred in Northern Italy around Fibonacci's lifetime—the creation of long-term government debt. Venice and Genoa both began to issue forced loans to their wealthy citizens in the late twelfth century. In 1171, for example, Venice issued a mandatory 5% loan to all citizens of the Republic to finance the construction of a fleet to fight Byzantium.[24] Out of this and successive loans called *prestiti* a regular market for government debt developed in Venice by the mid-thirteenth century. The consolidation of the Venetian debt by decree in 1262 institutionalized the funding of municipal debt, and gave birth to the now widespread modern practice of financing the government through a national debt.

[22] Ashtor, 1983. Cited in Kohn, 1999, "Bills of Exchange and the Money Market to 1600," working paper, Dartmouth College. The latter is an up-to-date and excellent survey of bills of exchange.

[23] Sivéry, 1983. Quoted in Spufford, 2002.

[24] Pezzolo, 2005.

The roots of this debt revolution can be found in the financial records of Pisa itself. Pisa in Leonardo's time regularly made long-term debt commitments to creditors, promising the perpetual flow of revenues from specific sources to amortize a debt. Favier reports just such a contract with a creditor of the commune in 1173.[25] Historian David Herlihy points out that the public finances of the commune became increasingly complex through the course of the thirteenth century. Pisan governors—*podesta*—began to raise money through capitalization of taxation and monopoly rights.[26] The economic decision to alienate a stream of future cash flows in return for a current lump-sum payment is essentially a present-value consideration. There is no doubt that the mathematical techniques in *Liber abbaci* would have allowed the *podesta* to quantify these decisions in a manner not feasible before Fibonacci's work. In fact, the annuity settled on Fibonacci late in his life for service to the state suggests that he may have had a hand in advising on precisely these financial contracts.

The technology of present value in *Liber abbaci* would also have served the analysis of private commercial transactions. Consider, for example, the extraordinary futures contracts quoted in neighboring Pistoia from at least the beginning of the thirteenth century. From 1201 to 1210, contracts for regular, ongoing deliveries of a *staio* of wheat sold for a median price of 2.80 lira—a period in which, according to Herlihy, the "spot" price of a staio sold for 1/4 of a lira. Without a mathematical method for discounting future deliveries, how did market participants determine the present value of such contracts?[27] How were speculators able to arbitrage the relative prices of farmland and futures contracts? Indeed, did the capacity

---

[25]   Favier, 1971, p. 283.

[26]   Herlihy, 1958. Chapter 6 describes these new contracts. Pezzolo, in Goetzmann and Rouwenhorst (2005), cites examples of similar sales of income rights in Genoa in 1152 and Venice in 1164.

[27]   Herlihy, 1967, pp. 138–39.

to discount future values exist in Northern Italy before Fibonacci or did his mathematics make futures and complex fixed-income contracts possible?

Overall, it is difficult to disentangle the causal relationship between the commercial problems of the day and the mathematical methods for solving them. Almost certainly, as is true today, there was a dynamic relationship between the analytical framework and the market. Mathematicians then, as now, must have been motivated by problems in the world around them, and in turn, their solutions engendered new advances in financial engineering.

Though Leonardo would soon be forgotten, his commercial legacy was already underway during his lifetime.

CHAPTER 16

Reflections in a Medieval Mirror

On May 12, 1983, I opened my copy of the *Guardian*, the British national daily newspaper, and there, in the weekly science section, called *Futures*, was my first guest article.

THE BIGGEST PRIME NUMBER IN THE WORLD

A prime number is any whole number that can only be divided (without recourse to remainders or fractions) by the numbers 1 and itself. (For example, 2, 3, 5, 7, 11 are all primes.) Although the prime numbers have been studied by mathematicians (both professional and recreational) since ancient times, it is only over the last few years that they have attracted interest from other quarters.

As reported on this page some weeks ago, recent developments in cryptology (the science of making and breaking secret codes) have tended to involve more and more aspects of the branch of mathematics known as number theory, and in particular the properties of prime numbers. Not surprisingly, therefore, large communications and data organizations such as IBM and the Bell Telephone company now provide extensive funding for research into prime numbers, and it is widely believed (though not, of course, acknowledged) that less academic agencies such as the CIA are also highly involved in such matters. So it is unlikely that only a handful of ivory-towered mathematicians will show interest in the recent announcement that a new prime has been discovered, a prime immeasurably larger than any known beforehand.

This number is so large that it would be pointless trying to represent it in the way numbers are usually expressed, using a string of the digits 0 to 9. If the editor of the *Guardian* were to decide to print this number in the normal way, using regular sized type, with no headlines, advertisements, or pictures in the way, the number would take up just over three and a half pages of the newspaper.

Fortunately, mathematicians have a special notation for describing numbers of this magnitude. Using this notation, the number in question looks quite tame. It is $2^{86,243}-1$.

That this number is prime was discovered by David Slowinski of the United States. As you might imagine, he had more than a £5 pocket calculator to help him with his calculation. In fact, he made use of the world's most powerful computer, the giant Cray-1 machine at the Cray Research Laboratories. Even with this incredible computing power, it took the machine 1 hour 3 minutes and 22 seconds simply to check that the above number is indeed a prime. Months of computing were required to find this number in the first place.

It is not hard to explain what the notation used above means. To obtain Slowinski's number you take the number 2 and multiply it by itself 86,243 times, and then, as a final fillip, you subtract 1. The result is a number with precisely 25,962 digits when written out in the normal way.

How can we begin to comprehend the size of such a monster? To get some idea, let's look at the apparently insignificant number 2^{64}. This can be visualized as follows. Imagine an ordinary chessboard. If we number the squares on this chessboard starting in the top left-hand corner and proceed row by row down to the bottom right-hand corner, using the numbers 1,2,3 and so on, the last square we number will get the number 64.

Now imagine that we start putting ten-pence pieces on the squares of the chessboard. On square number 1 we put 2 10p pieces, on square 2 we put 4, on square 3 we put 8, and so on, on each successive square putting exactly twice as many coins as on the previous one.

On the last square we will form a pile of exactly 2^{64} ten-pence pieces. How high do you think this pile will be? Six feet? Fifty feet? More? Wait for it. The pile will be about 37 million million kilometers high! So the pile would stretch way beyond the moon (a mere 400,000 kilometers away) and the sun (150 million kilometers from Earth), and in fact would reach the nearest star, Proxima Centauri.

And that is only for the 2^{64}. To reach Slowinski's new prime you have to double up the pile of coins an additional 86,179 times. You would have left the entire universe long before you got there.

Why should anyone be interested in such huge numbers? There are various answers to such a question. To the mathematician, the way the prime numbers are distributed amongst all the numbers is an extremely interesting question in its own right. No one can say just where the next prime number will be. With small numbers, there appear to be lots of primes about, for instance, of the numbers less than 25 the numbers 2,3,5,7,11,13,17,19,23 are all primes. But as soon as you start looking at much larger numbers, the primes become much less frequent, though they do not appear to follow any particular pattern.

Aside from this perhaps esoteric interest, like almost all pure research there are various useful offshoots from the work. For instance, simply to get the computer to handle a number with 86,243 binary digits, like Slowinski's, an entire discipline of computer science known as multi-precision arithmetic has had to be developed, and you can bet your last prime that the CIA (among others) are interested in that.

Within a few hours, I got a phone call from the BBC saying they wanted to air a piece about the discovery on their early evening television news program, which was geared toward younger viewers. A short time after that I received another call from the BBC asking if I would help them to record a segment for the popular children's television show *Record Breakers*.

I did not know it at the time, but my second career as a math popularizer had just been launched. In addition to the interest from the BBC, a number of senior people at the *Guardian*, most of them politics, history, and literature majors, had commented favorably on the piece.

Reading my article today—as I just did copying the text into this manuscript—it seems fairly run-of-the-mill. But things were different back then. To the best of my knowledge, my piece was the first time any British daily newspaper had published a news story on a mathematical discovery, written by a professional mathematician.

It was also, I think, the first time the hunt for record prime numbers made the leap from mathematics journals and magazines into the daily press. (In the ensuing years, such stories have appeared with some regularity, as more and more new record primes have been discovered. I think I wrote about two, possibly three, of those further discoveries, but no more.)

At the time, however, all I knew was that I had written an article, it got published, and BBC Television immediately ran a cover story and was going to record a second.

It was a few days later, when the *Guardian* Science Editor called me up to say he would be interested in receiving additional pieces from me, that I started to think in terms of writing for them on at least an occasional basis. That was when I realized I had a side career in popular mathematics writing.

It was that secondary career activity that would eventually lead to my fascination with Fibonacci, as I referred to him before I "got to know him" and decided he deserved to be called by his real name, not a nineteenth-century *nom de plume*, invented by someone else.

But there was something else about the reaction my first article produced, that I observed over and over again throughout my mathematics outreach career. There is a huge fascination with numbers and mathematics—*provided it is presented in an interesting fashion.*

Over my ensuing decades of mathematics outreach activities, starting with the twice-monthly column in the *Guardian* that followed soon after my initial Record Prime Number piece, I discovered time and again just how deeply attached people are to numbers. Often, this attachment manifests itself in numerology, with books like *The Bible Code*[1] becoming international best sellers.

---

[1] No, I am not going to give a reference. It's junk science that has been amply demonstrated as such. Its central claim that there are numerically encoded secret messages for Mankind in the Holy Book is spurious, and the method described will produce (not uncover!) "hidden messages" in any book roughly the same size as the Bible.

I also found it when I would write about certain genuine mathematical topics, none more so than whenever I covered the Monty Hall Problem.[2] This tricky probability conundrum baffles many people on first encounter, mathematicians among them, occasionally famous ones. The aspect that I find most interesting is not what people think or say, but the degree of passion they bring to the discussion. When a simple problem about the arithmetic behind a 1950s TV quiz show generates passionate argument, you'd better admit that numbers and arithmetic are a significant part of human cognition, whatever folk may say about the degree to which they are "numbers people."

Then there are the golden ratio and the related Fibonacci sequence that came up in chapter 10. My attempts over the years to separate for my readers and my public talks audiences the correct claims about the golden ratio from the fanciful inventions that these days live on the Internet, inevitably produce vehement and occasionally abusive critiques. People really care. About a single number!

Human beings, it seems, want to believe that their lives and destiny are somehow governed by numbers and equations. To be sure, there is a respect in which they are; namely, we can write equations that *describe* various aspects of our lives. But descriptive is (in general) not prescriptive. (What's more, as it happens, outside of botany, the golden ratio rarely occurs in either a descriptive or prescriptive fashion.)

Again, when I worked as a math consultant on the television crime series *NUMB3RS* from 2005 to 2008, one of the main messages the producers kept getting from their audience research groups was "Include more of the math." There too, I saw that ordinary citizens have a reverence for—and interest in—numbers, even if they profess to be bad at math.

---

[2]   Start with the Wikipedia entry. Please don't write to me about it. On this one particular topic, I won't read what you say—I have seen it all before—and I won't reply. Other topics, yes. But not this one.

Yet, despite this deep attachment to numbers, particularly in the Western nations people take pride in declaring that they are "not a numbers person." Why this seeming dichotomy?

Almost certainly, some of the negativity I observe toward mathematics is a consequence of bad experiences in poorly taught math classes at school. It is a natural defense mechanism, when faced with a bad math experience, to convince yourself that you do not think mathematics is important, and have in fact "chosen" not to do it—or even worse, that you are not capable of doing it.

But there is also social support for that mind-set, with a long history going back at least as far as Ancient Greece, from which point we in the Western nations inherit much of our culture and our intellectual attitudes.

As many of us learn in school, ancient Greek society revered intellectual activity pursued for its own sake, but looked down upon artisanal labor. Mathematics, of course, covers both camps, with "pure" intellectual subjects such as geometry and number theory in the "artistic" category and arithmetic—particularly commercial arithmetic—in the artisanal.

When I was growing up in England, this societal attitude was extremely strong, with bright kids like me constantly being urged to pursue the arts (which, in true adherence to the revered culture of the ancient Greeks, were declared to include the works of Euclid) and not waste our talents on the more "mundane," utilitarian sciences.

The famous English number theorist Godfrey Harold (G. H.) Hardy, in his celebrated book *A Mathematician's Apology*, published in 1940, wrote, with evident pride:

> I have never done anything "useful." No discovery of mine has made, or is likely to make, directly or indirectly, for good or ill, the least difference to the amenity of the world.

A bit later, in a lecture delivered at Cambridge University that was subsequently expanded into a book, the English scientist and novelist Charles Percy (C. P.) Snow talked about the "Two Cultures" in British intellectual life—the arts on the one side, the sciences on the other.[3]

The point Snow was making was that the arts versus sciences cultural divide, with its inherent ranking of the former above the latter, would be highly damaging to any subscribing nation in what was very clearly an emerging new era dominated by science and technology.

Emphasizing the degree to which Snow was right on target in that regard, in the 1980s, mathematicians took some of G. H. Hardy's beloved number theory and used it to forge the encryption system that underpins Internet communication, making possible online commerce, modern warfare, and widespread government surveillance, and refuting Hardy's statement about its lack of utility.

The fact is, the artistic–scientific divide always was artificial. As products of the human mind, mathematics and the sciences always were, and surely always will be, every bit as creative and aesthetically rich as any of our arts or literature. The problem lies not in the different flavors of human thought, but in how the different disciplines are sometimes taught.

As I complete this account of the writing of my first fully fledged history book, reflecting on the ten-year process of bringing it to life, I cannot but smile at the irony of my having found history the most boring of all subjects at school.[4] Yet, there is no mystery

---

[3]   Snow, 1959.

[4]   In 2010, I did publish a short book about the early development of modern probability theory, *The Unfinished Game: Pascal, Fermat, and the Seventeenth-Century Letter that Made the World Modern* (Basic Books), but by then my Leonardo project was nearing completion, and besides, that book required little in the way of historical research.

as to how that came about. The school history curriculum I was subjected to in 1960s England was a test-driven ordeal of memorizing names and dates of all the British kings, queens, and the wars and battles they fought. (There are a lot of them—monarchs, wars, and battles!) Facts to be studied and retained long enough to pass the test. The trick to doing well on an examination essay question was to cram in (from memory) as many factual references, mainly names and dates, as possible. My fellow students and I all got hold of cheap cramming booklets that were nothing more than lists of the key facts to memorize to pass the exam·

I obtained similar cramming books for my English Literature classes. One particularly useful one reduced each of Shakespeare's plays to a two- or three-page list of the key facts required to pass the test. (Needless to say, it was only long after I had graduated from high school that I actually began to find Shakespeare enjoyable.)

By good fortune, I had very good mathematics and science teachers in high school, and that, coupled with my early desire to "go into space science," meant that my school mathematics and science education continually stimulated my interests. I never found the subjects easy; to this day I find doing mathematics hard. That, to my mind, is what makes it so interesting. Anyone who thinks that doing math is easy does not know what math really is—just as the middle-school me did not really know there is more to history and literature than simply memorizing facts. But though I always found mathematics hard, I never lost interest in what was self-evidently (to me, and I suspect to anyone who overcomes any negative effects of bad school mathematics classes) a towering edifice of collective human thought over several thousand years.

To a modern reader who picks up *Liber abbaci*, the book is a laborious sequence of arithmetical routines for performing computations, each one illustrated with copious worked examples. Textbook recipes for performing calculations, they seem as far removed from "creative, human intellectual activity" as you can imagine.

But take a step back. Instead of getting lost (as many do) in the devilish details (of which there is no shortage), look at what is going on. More precisely, look at what Leonardo and the "numbers men" who went before him were trying to do.

We humans evolved to be social creatures who meet our needs and fulfill our desires by cooperating with others. ("Cooperation" can include competing, of course. Two people playing chess may be competing, perhaps fiercely, but they do so only by cooperating in playing the game, both players following its rules.) In the modern parlance of systems theory, we form an interactive system.

But that system is not one of random interactions, like the molecules of gas bouncing off one another in a container. Sure, there are some random interactions in the human system, but over millions of years of evolution by natural selection, we have developed ways of acting and ways of interacting that produce patterns of interactions that benefit our survival.

In particular, we have learned to be more specialized, to focus on different activities, relying on the skills and expertise developed by others to help us meet our needs and fulfill our desires.

The advantage of this endless trend to ever more specialization is that it enables each one of us to focus more and more on the things that interest us most. Today, hardly any of us know how to grow vegetables or how to rear and keep farm animals. We leave it to others to ensure we have food. Just as we leave it to others to provide us with pleasing—and sometimes challenging—works of art, to stimulate and entertain us with novels and movies, to soothe or excite us with music, to cure us when we are ill, to build and maintain our transportation devices, to provide us with various communication and information technologies, and to help us understand ourselves and the world we live in from many different perspectives (the natural and human sciences).

The price we pay for being able to indulge our own particular interests is that we are often unable to see beneath the surface of the activities of others. When all we see of others are simple input-output behaviors (e.g., we take our car to the repair shop and two days later we pick it up with the problem diagnosed and fixed), it is easy to think that it requires little by way of human expertise, ingenuity, or creativity. But that is rarely the case. We are, after all, talking about people. Human beings, just like us. Yes, we are highly attuned to noticing the differences between us, but we all have far more in common with one another than we do differences.

To be sure, much of what we do becomes routinized and performed repetitively. But that does not make us mindless robots. To classify the actions of a fellow human as mindless or boring is to view that individual through our own perspective. And to rank people, with one being "higher" than the other (say, in terms of "creativity"), is simply egocentric, regardless of whether we rank ourselves as higher, like the literary types C. P. Snow railed against, or lower, like the many who claim they cannot do math. Having the relative freedom to live our lives as we choose is possible only because we live in the evolved, complex, interactive system we call human society.

There is little evident creativity in the way I follow a recipe to prepare a cooked meal. When I want a really good meal, I go to a restaurant. In so doing, I do not say to myself, "I'll go to Chez Panisse because Chef Alice Waters is much better at following a recipe than I am." I do so because I know she is simply a very good cook. She has creative expertise.

Similarly (and deliberately getting away from a domain as prone to "X-snobbery" as when X = food), I have in the past occasionally found myself unable to successfully mount a new tubeless tire on my road bike, despite having read online manuals and watched YouTube videos demonstrating how to do it, and have ended up taking the wheel and tire to my local bike shop. Nothing, surely, can look any less requiring of skill, expertise, or creativity than putting a tire on a bicycle wheel. I know. Besides watching those YouTube videos, I would stand alongside the bike shop mechanic as he quickly and effortlessly put the tire onto the wheel. Simple, no?

Clearly, what he was doing—and *experiencing*—was very different from what I was doing and experiencing when I tried and failed. I *knew* it simply could not be a trivial task, since I had been unable to do it. But what was I not seeing?

In this case, having (for good reasons) switched to road tubeless wheels, I could not leave that question unanswered. Because I often ride my bike alone in remote areas of open countryside, it was important to me to acquire at least some of the skill the bike shop mechanic had in case I had a puncture. So I kept persevering. It was not easy. But eventually, after many attempts and much reflection on what was required, I learned how to do it. My big breakthrough occurred the one time when the mechanic, holding the wheel horizontally pressed to his stomach, while manipulating the tire with both hands, told me what he was really doing. "You have to think of the tire as alive," he said. "It wants to be sitting firmly on the rim [that, after all, is what it was designed for], but it is not very disciplined. Like a small child, it moves around and resists your attempts to force it. You have to understand it, and be aware, through your hands, of what it is doing. Work with it—be constantly aware of what it is trying to do—so you both get what you want: the tire gets onto the wheel and you can inflate it and get back on your bike."

Fanciful? Maybe. But it worked. And it continues to work. As a result, not only can I now change my tubeless tires, it has for me become "mindless and automatic," as effortless as Picasso drawing a simple doodle on a restaurant napkin to pay the bill for his meal. It took many years for Picasso to learn to draw the way he did (and for the marketplace to assign high value to his work), but that does not mean his work was not creative; rather, he simply routinized part of it. When I watch a film of him at work, I see superficially how he creates, but I do not see his canvas as he does, and I could not draw as he does. Likewise, my skill in fitting a tubeless tire is a result of my now seeing and understanding what earlier had been opaque. (I admit that it is far easier to learn to mount a tubeless tire on a road bike wheel than to draw like Picasso. But it is possible to learn from the analogy.)

Getting back to Leonardo and *Liber abbaci*, he was part of many generations of traders, bankers, and mathematicians who used their creative intellects to develop better and ever more powerful methods for handling the numerical calculations that became steadily more essential for fueling the network of inter-human interactions of various kinds. To those pioneers, numbers were (and of this I am absolutely sure) no less "alive" and "with character" than the bicycle tire was to my bike shop mechanic (and now to me) and Picasso's canvas was to him. They were developing, and in Leonardo's case finding, creative ways to explain to others the methods and procedures that they could use in their daily lives—including their recreational time, as evidenced by some of the more fanciful "fun problems" Leonardo included in his book.

One of the problems that bedevils mathematics is, perhaps ironically, that math is so darned relevant to so many walks of life. That means that everyone should receive a mathematics education. They should know something about mathematics in order to be a fully functioning, contributing member of society.

But that should not be taken to mean that the focus should be all on "useful math" (whatever that might mean—and G. H. Hardy would have gotten that wrong). As I wrote in the introduction to my book, *Introduction to Mathematical Thinking*:

> Education is not solely about the acquisition of specific tools to use in a subsequent career. As one of the greatest creations of human civilization, mathematics should be taught alongside science, literature, history, and art in order to pass along the jewels of our culture from one generation to the next. We humans are far more than the jobs we do and the careers we pursue. Education is a preparation for life, and only part of that is the mastery of specific work skills.[5]

Leonardo wrote *Liber abbaci*, and even more so his "book for merchants," for the society of the time. It was his living, creative contribution to human social activity. What may appear to us today as dull, repetitive, and boring, does so simply as a consequence of changed circumstances—changes brought about in substantial part by those very books. The fame that *Liber abbaci* brought to Leonardo throughout Italy in his lifetime should indicate that the book offered far more than we can see in it today. My recognition that there was a strong similarity between Leonardo's evident enthusiasm and zeal in his work as a producer of books and that of Steve Jobs in producing the Apple I and II, the Macintosh, the iPhone, and the iPad is what led me to write my electronic supplement to *The Man of Numbers*, paralleling the two men's careers, as I summarized earlier in chapter 14.[6]

In today's society, Jobs's creativity in the technology arena is never doubted. Whatever else we may think about him, we must

---

[5] Devlin, 2012, p. 8.
[6] *Leo and Steve: The Young Genius Who Beat Apple to Market by 800 Years*, 2011.

regard him as a genius of the twentieth century. That, I suggest, is the way we should view Leonardo within the thirteenth century. His books are among the great cultural and intellectual creative artifacts of human history. They are "insanely great," as author Steven Levy wrote of Jobs's creation of the Macintosh computer.[7] *Liber abbaci* is Leonardo's *Mona Lisa*, his *David*, his *Ninth Symphony*, his Notre Dame Cathedral, his Golden Gate Bridge, and yes, his iPad. As such, Leonardo's work is an important part of human cultural history.

In one respect, Leonardo's work is perhaps even more important than that of many others. For, whereas not everyone appreciates Da Vinci, Michelangelo, or Beethoven, or visits Notre Dame, or crosses the Golden Gate Bridge, and while not everyone walks round with a smartphone in their pocket,[8] we all of us carry around in our minds the elements of basic arithmetic that Leonardo helped put there.

---

[7] Levy, 1994.
[8] This one may not remain true for much longer, which in view of chapter 14 actually strengthens my point.

APPENDIX

Guide to the Chapters of Liber abbaci

The exact titles of the 15 chapters of *Liber abbaci* vary from manuscript to manuscript, suggesting that the scribes who made copies felt free to make what they thought were clarifying improvements. But, give or take the actual wording, Leonardo divided up *Liber abbaci* as follows. (The page counts are for Sigler's English-language translation.)

1. *On the recognition of the nine Indian figures and how all numbers are written with them; and how the numbers must be held in the hands, and on the introduction to calculations*
   Following the two-page Dedication and Prologue, Chapter 1 occupies 6 pages in Sigler's translation, and describes how to write—and read—whole numbers in the Hindus' decimal system. For large numbers, the numerals are grouped in threes to facilitate reading. Leonardo also describes a system for computing with the fingers of the hands, illustrated by elaborately drawn diagrams in manuscript copies of *Liber abbaci*. This system made it possible to

perform calculations more efficiently, and was widely used during the medieval period. He also provided addition and multiplication tables to be referred to—or, far better, memorized—in order to facilitate computations.

With the reading and writing rules out of the way, Leonardo devotes Chapters 2 to 15 to the rules and methods of arithmetic.

2. *On the multiplication of whole numbers*
This is a 16-page "how-to" manual. The approach differs little from the one used today to teach children how to multiply two whole numbers together. Leonardo begins with the multiplication of pairs of two-digit numbers and of multi-digit numbers by a one-place number, and then works up to more complicated examples. He describes various methods for checking the answers. One of them, "casting out nines," was just going out of use in the UK when I was taught arithmetic there in the early 1950s, and in today's era of cheap ubiquitous electronic calculators there is little reason to spend time explaining it—so I won't.

3. *On the addition of them, one to the other*
Chapter 3 is short, with just 5 pages of instructions. Today's reader probably finds it strange that he treats addition after multiplication, but there it is. Giving a hint of things to come, he describes a procedure for keeping expenses in a table with columns for librae, soldi, and denari.

4. *On the subtraction of lesser numbers from greater numbers*
This, the shortest chapter in the book, occupies a mere 3 pages of Sigler's translation. The title tells you all.

5. *On the divisions of integral numbers*[1]

Chapter 5 focuses on divisions by small numbers and on simple fractions. It is considerably longer, the longest so far, with 28 pages of instructions. Division and fractions are hard and require more explanation. Today's readers familiar with the standard "long-division algorithm," once a staple of school mathematics education but now an optional extra in many countries, will, *with a bit of effort*, recognize an old friend.

6. *On the multiplication of integral numbers with fractions*

The topic is what are today called mixed numbers, numbers that comprise both a whole number and a fractional part. Leonardo explains that you calculate with them by first changing them to fractional form (what we would today call "improper fractions"), computing with them, and then converting the answer back to mixed form. This chapter takes up 22 pages.

7. *On the addition and subtraction and division of numbers with fractions and the reduction of several parts to a single part*

Leonardo fills 28 pages showing how to combine everything that has been learned so far.

8. *On finding the value of merchandise by the Principal Method*

With Chapter 8 we get our first real dose of practical mathematics, in the form of 51 pages of worked examples on the value of merchandise, using what we would today call reasoning by proportions—the math we use to check the best deal in the supermarket. For example, Leonardo

---

[1] An *integral number*, or *integer*, is a technical term for a whole number, positive, negative, or zero.

asks us: if 2 pounds of barley cost 5 soldi, how much do 7 pounds cost? He then proceeds to show us how to work out the answer. He makes use of simple diagrams of proportion, which he calls the "method of negotiation." Examples include monetary exchange, the sale of goods by weight, and the sale of cloth, pepper, cheese, canes, and bales.

9. *On the barter of merchandise and similar things*

Leonardo presents another 33 pages (in the Sigler translation) of worked practical examples extending the discussion from the previous chapter. There are problems on the barter of common things, on the sale and purchase of money, on horses that eat barley in a certain number of days, on men who plant trees, and men who eat corn.

10. *On companies and their members*

These 14 pages provide worked examples on investments and profits of companies and their members, showing how to decide who should be paid what.

11. *On the alloying of monies*

Chapter 11 occupies 31 pages in Sigler's English-language translation. The need for the methods Leonardo describes in this chapter was considerable. At that time, Italy had the highest concentration of different currencies in the world, with 28 different cities issuing coins during the course of the Middle Ages, 7 in Tuscany alone. Their relative value and the metallic composition of their coinage varied considerably, both from one city to the next and over time. This state of affairs certainly meant good business for money changers—and *Liber abbaci* provided plenty of examples on problems of that nature. But in addition, with governments regularly re-valuing their currencies, gold and silver coins provided a more stable base, and since most silver coinage of the time was alloyed with copper, problems of minting and alloying of money were important.

12. *On the solutions to many posed problems*

This enormous chapter fills a staggering 186 pages with miscellaneous worked examples. Its primary focus is algebra. Not the symbolic reasoning we associate with the word today, but rather "algebraic reasoning," expressed in ordinary language (and often referred to as "rhetorical algebra"). Much of Leonardo's focus is on applications of what is generally known as the "method of false position," which he refers to as the "tree method." This is a procedure used to solve problems equivalent (in modern terms) to a simple linear equation of the type $Ax = B$. The solver first picks an approximate answer and then reasons to adjust it to give the correct solution. Many of the problems Leonardo looks at he called "tree problems," which is why he speaks of solving them by the "method of trees." This is a class of problems he named after a particular puzzle he introduces in the chapter, where you want to know the total length of a tree when you are given the proportion that lies beneath the ground.

He also shows how to solve the same kinds of problems using what he called the "direct method" (*regula recta*), where you begin by calling the sought-after quantity a "thing" (*res*), and then form an equation (expressed in words), which is then solved step-by-step to give the answer. Expressed symbolically, this is precisely the modern algebraic method. It was known to the Arab scholars, and was described by the mathematician al-Khwārizmī around 830 in the book from whose title the modern word "algebra" stems. I say more about al-Khwārizmī in chapter 11.

Many of the problems Leonardo presents are of a financial nature, providing the businessmen of the thirteenth century and thereafter with some extremely powerful tools that helped to revolutionize European trade and commerce. (See my chapter 12.)

13. *On the method elchataym and how with it nearly all problems of mathematics are solved*

In modern terminology, *elchataym* is a rule, known also as "double false position," used to solve one or more linear equations. The word "elchataym" is Leonardo's Latin transliteration of the Arabic *al-khata'ayn*, which means "the two errors." The name reflects the fact that you start with two approximations to the sought-after answer, one too low, the other too high, and then reason to adjust both until the correct answer is arrived at. It can be used to solve linear equations not only of the form $Ax = B$, for which single false position can be used, but also the more general form $Ax + B = C$. This chapter provides 41 pages of worked examples. Leonardo formulates the problems in several ingenious ways, in terms of snakes, four-legged animals, eggs, business ventures, ships, vats full of liquid which empty through holes, how a group of men should divide the proceeds when they find a purse or purses, subject to various conditions, how a group of men should each contribute to the cost of buying a horse, again under various conditions, as well as some in pure number terms.

14. *On finding square and cubic roots, and on the multiplication, division, and subtraction of them, and on the treatment of binomials and apotomes and their roots*

Leonardo's penultimate chapter offers 42 pages of worked examples. His main focus is on methods for handling roots. He uses the classifications given by Euclid in Book X of *Elements* for the sums and differences of unlike roots, namely binomials and apotomes.

"What are *apotomes*?" you ask. Recall that the discovery that $\sqrt{2}$ is irrational led the ancient Greeks to a study of what they called "incommensurable magnitudes." Euclid's term for a sum of two incommensurables, such as $\sqrt{2} + 1$, was *binomial* (a "two-name" magnitude), and a difference, such as $\sqrt{2-1}$, he called an *apotome*. Handling incommensurables by means of what we would now regard as algebraic expressions was a common feature of Greek and medieval mathematics, and was the same way they dealt with algebraic binomials such as "two things less a dirham." This is one reason why many medieval Italian treatments of algebra began with a long chapter on binomials and apotomes.

In terms of mathematical content, Chapter 14 is little more than a collection of known methods and results, and Leonardo presents nothing significant not already found in *Elements*.

15. *On pertinent geometric rules and on problems of algebra and almuchabala*

This final chapter occupies 85 pages, again filled with worked examples. What is *almuchabala*? In modern terms it corresponds to manipulating the two sides of an equation while keeping it balanced. Once you know that, the chapter title says it all. Leonardo's approach differs little from that found in al-Khwārizmī's earlier book on algebra.

The origin of our word *algebra* is interesting. What we now refer to as "algebra" was called *al-jabr wa'l-muqabala* in Arabic, or sometimes just *al-jabr* for short. The two words *al-jabr* and *al-muqabala* refer to two steps in the simpli-fication of equations. The former means restoration or completion. In an equation like "ten less a thing equals five things," the "ten less a thing" was thought of as a deficient "ten" that needs to be restored. The Arabic mathematicians would write, "So restore the ten by the thing and add it to the five things," to get the equation "ten equals six things." The other term, *al-muqabala*, means confrontation. In an equation like "ten and two things equals six things," you confront the two things with the six things, which entails taking their difference, to get the equation "ten equals four things."

Sigler writes "algebra and almuchabala" for Leonardo's "algebre et almuchabale," which is a transliteration of the Arabic *al-jabr wa'l-muqabala*, but this entire expression really just means "algebra." Jeffrey Oaks, the contemporary expert in medieval algebra who I consulted when I was working on *The Man of Numbers*,[2] points out that Leonardo translated the two terms as "proportionem et restaua-tionem" (proportion and restoration), but in so doing gets them the wrong way round. "Restarare" was the common translation of *al-jabr*, while "proportionem" is a poor ren-dering of *al-muqabala*. In fact, Oaks adds, Leonardo used the term "restarare" in a way unlike the Arabic *al-jabr*, showing that he did not understand the use of the term.

---

[2]   Private communication, 2010.

Bibliography

Arrighi, G., ed. 1964. Paolo Dell'Abaco, *Trattato d'aritmetica*. Pisa: Domus Galilaeana.

———, ed. 1967. Antonio de' Mazzinghi, *Trattato di Fioretti*. Siena: Dmous Galilaeana.

———, ed. 1973. *Libro d'abaco*. Lucca: Cassa di Risparmio di Lucca.

———, ed. 1987. Paolo Gherardi, *Opera mathematica: Libro di ragioni-Liber abaci*. Lucca: Pacini-Fazzi.

Ashtor, Eliyahu. 1983. *Levant Trade in the Later Middle Ages*. Princeton, NJ: Princeton University Press.

Bernardini, Rodolfo. 1977. "Leonardo Fibonacci nella iconografia e nei marmi" [Leonardo Fibonacci in iconography and in marbles], *Pisa Economica*, (n.1), *Pisa Camera di Commercio, Industria, Artigianato* (Chamber of Commerce), pp. 36–39.

Devlin, Keith. 2000. *The Math Gene: How Mathematical Thinking Evolved and Why Numbers Are Like Gossip*. New York: Basic Books.

———. 2011. *The Man of Numbers: Fibonacci's Arithmetic Revolution*. New York: Walker Books.

———. 2011. *Leo and Steve: The Young Genius Who Beat Apple to Market by 800 Years*. Amazon Kindle eBook original.

———. 2012. *Introduction to Mathematical Thinking*. Amazon Kindle eBook original.

Favier, Jean. 1971. *Finance and Fiscalité au Bas Moyen Âge*. Paris: Société D'Édition D'Enseignement Supérieur.

Fisher, Irving. 1930. *The Theory of Interest*. New York: Macmillan.

Franci, Rafaella. 2003. "Leonardo Pisano e la Trattatistica dell'Abaco in Italia nei Secoli XIV e XV," *Bollettino di Storia delle Scienze Mathematiche*, Intituti Editoriali e Poligrafici Internazionali, 23, Facs. 2, pp. 33–54.

Gies, F., and J. Gies. 1969. *Leonardo of Pisa and the New Mathematics of the Middle Ages*. N.p.: Crowell Press.

Goetzmann, William N. 2004. *Fibonacci and the Financial Revolution*. National Bureau of Economic Research, Working Paper 10352, March.

———. 2005. *Fibonacci and the Financial Revolution*. In Goetzmann and Rouwenhorst, eds., *Origins of Value: The Financial Innovations that Created Modern Capital Markets*. New York: Oxford University Press, 2005, pp. 123–143. (A revised version of Goetzmann 2004.)

Goetzmann, William N., and K. Geert Rouwenhorst, eds. 2005. *Origins of Value: The Financial Innovations that Created Modern Capital Markets*. New York: Oxford University Press.

Graham, John R., and Harvey Campbell. 2001. "The Theory and Practice of Corporate Finance: Evidence from the Field," *Journal of Financial Economics*, 60, no. 2–3, pp. 187–243.

Grimm, Richard E. 1973. "The Autobiography of Leonardo Pisano," *Fibonacci Quarterly*, 11, no. 1 (February), pp. 99–104.

Hardy, G. H. 1940. *A Mathematician's Apology*. Cambridge, UK: Cambridge University Press.

Heefler, Albrecht. 2007. "The Abbacus Tradition: The Missing Link between Arabic and Early Symbolic Algebra." *Proceedings of the International Seminar on the History of Mathematics*, New Delhi, India.

Herlihy, David. 1958. *Pisa in the Early Renaissance: A Study of Urban Growth*. New Haven, CT: Yale University Press.

———. 1967. *Medieval and Renaissance Pistoia*. New Haven, CT: Yale University Press.

Hughes, Barnabas. 2004. "Fibonacci, Teacher of Algebra: An Analysis of Chapter 15.3 of Liber Abbaci," *Mediaeval Studies*, 64, pp. 313–61.

———. 2008. *Fibonacci's De Practica Geometriae*. New York: Springer-Verlag.

Kohn, Meir. 1999. "Bills of Exchange and the Money Market to 1600." Dartmouth College, Working Paper. http://www.dartmouth.edu/~mkohn/99-04.pdf.

Levy, Steven. 1994. *Insanely Great: The Life and Times of Macintosh, the Computer that Changed Everything*. New York: Penguin Books.

Livio, Mario. 2002. *The Golden Ratio*. New York: Broadway Books.

Markowsky, George. 1992. "Misconceptions About the Golden Ratio," *College Mathematics Journal*, 23, no. 1 (January), pp. 2–19.

Munro, John H. 2003. "The Medieval Origins of the Financial Revolution: Usury, Rentes, and Negotiability," *International History Review*, 25, no. 3 (September), pp. 505–62.

Pezzolo, Luciano. 2005. "Bonds and Government Debt in Italian City States: 1250–1650." In Goetzmann and Rouwenhorst, eds., *Origins of Value:*

The Financial Innovations that Created Modern Capital Markets. New York: Oxford University Press.

Pryor, John H. 1977. "The Origins of the Commenda Contract," *Speculum*, 52, no. 1 (January), pp. 5–37.

Rosen, Frederic. 1831. *The Algebra of Mohammed Ben Musa.* London: J. Murray.

Sigler, Laurence. 1987. *The Book of Squares: An Annotated Translation into Modern English.* New York: Academic Press.

———. 2002. *Fibonacci's Liber Abaci: A Translation into Modern English of Leonardo Pisano's Book of Calculation.* New York: Springer Verlag.

Sivéry, Gérard. 1983. "Mouvements de capitaux et taux d'interêt en occident au XIIIe siècle," *Annales Economies Societés Civilizations*, 38, p. 367.

Smith, D. E. 1951. *History of Mathematics*, Vol. 1. Mineola, NY: Dover.

Spufford, Peter. 1986. *Handbook of Medieval Exchange.* Offices of the Royal Historical Society, London.

———. 2002. *Power and Profit: The Merchant in Medieval Europe.* New York: Thames and Hudson.

Snow, Charles Percy. 1959. *The Two Cultures.* Cambridge, UK: Cambridge University Press.

van Egmond, Warren. 1980. *Practical Mathematics in the Italian Renaissance: A Catalog of Italian Abbacus Manuscripts and Printed Books to 1600.* Firenze: Istituto e Museo di Storia della Scienza; Monografia N. 4.

Index

abacus, 15, 30
abbacus book, 25
abbacus school, 26, 80
abbacus teacher, 26
abbacus vs. abacus, 15
"Abbacus Eve," 28
Adams, Greg, 98, 100–114
"algebra," origin of word, 32, 78,
    234–35
algebra, nature and origins of,
    32–33
algorithm, 31, 78
Apple Macintosh, 182–91, 226
"Arab," 18
Arrighi, Gino, 24

banking problems, 205–8
Barozzi, Guilo Cesare, 64–68, 86, 139
bicycle tires, tubeless, mounting,
    224–25
Bigholli, 153–55
birds, Fibonacci's problem of the,
    117–18, 136–37
Bonacci, Guilielmo, 18
Boncompagni, Baron Baldassore,
    67, 83, 86–87, 94, 107
"book for merchants," 29, 92,
    168–80

"book in a smaller manner," 28
Bugia, 18–19

calculus, 3
Camposanto, 48
Cantor's Continuum Problem, 9
Codex 2404, 170–80
Cohen, Paul, 9
commercial arithmetic, 24
Cossali, Pietro, 23, 34

De Practica Geometrie, 27–28, 91, 106,
    170, 172
denarii, 62
Devlin's Angle, 11
digit, 30
divine proportion,129
division of profits, 202–4

Egmond, Warren van, 24–25
Einstein, Albert, 21, 68
Elements, 12–13, 68
Erdös, Paul, 104
Euclid, 12, 68
exposition, mathematical, 12

Fell, Judith Sigler, 98–105
Fibonacci, 12–13

Fibonacci numbers, 126–33
"Fibonacci," origin of name 22–24
Fibonacci Quarterly, 37, 94, 133
Fibonacci sequence, 24, 36,
    115–26, 218
Fibonacci Society, 37, 133
Fibonacci statue, 46–55
filius bonaci, 22
financial mathematics, 192–212
finger arithmetic, 30, 89
Florence National Library, 82, 87,
    139, 146–48, 156–61, 176
Flos, 28
Florence manuscript of *Liber
    abbaci*, 82, 139
Franci, Rafaella, 16, 29, 39, 73–84,
    97, 140

Galileo, 21, 36
Gates, Bill, 189–91
Genoa, 21, 56
Giardino Scotto, 45–46, 54–55
Goetzmann, William, 17, 39, 169,
    192–212
golden ratio, 129–33, 218
Grimm, Richard, 94–96, 140
Guardian newspaper, 10, 213–17

Halmos, Paul, 105
Hardy, G. H., 219–20, 226
Harry Potter, 87, 134
Hindu-Arabic arithmetic, 12–13,
    30–31, 66, 70, 79–80, 88–93,
    190–97
Hughes, Barnabas, 83, 140, 165

interest rate problems, 205–8

Jobs, Steven, 183–191, 226–27

Khayyam, Omar, 78
al-Khwarizmi, 12, 31, 77–81, 196
Knuth, Donald, 111

Leaning Tower of Pisa, 43–44
Leonardo, life and dates, 22–24, 86,
    92–96, 163–66
Leonardo proclamation, 163–66
Levy, Steven, 182–87, 227
Liber abbaci, 13, 16,20, 66–67,
    80–96, 116–26, 136–37, 152–63,
    228–35
Liber minoris guise, 28, 170
Liber quadratorum, 28, 87, 99, 105–6
librae, 62
libri d'abbaco, 25
Libri, Gillaume, 22, 24, 128
Livero de l'abbecho, 170
Livorno, 57
long-term government debt,
    210–12
Lucas, Edouard, 24, 128
Lungarno Fibonacci, 36, 45
Lungarno Galileo, 36

maestri d'abbaco, 26
Man of Numbers, The, 16, 38–40, 71,
    90, 165, 167, 169, 235
Math Gene, The, 64, 74
mathematics interface, 189–90
mathematics, nature of, 10
"Math Guy," 11–12
Micromaths Column, 10
Montagno, Franco, 72–73
Monty Hall Problem, 218ff

National Public Radio, 11
Nesi, Jeannie, 158–61, 176
Newton, 21
notched bones, 70
NUMB3RS television series, 218
numbers, origin of, 70–71

Oaks, Jeffrey, 135

Pacioli, Luca, 23, 33–34
Paganucci, Giovanni, 48–50, 53

Pagli, Paoli, 76–84, 97
Palio, 142
personal computing revolution, 33
Piazza dei Miracoli, 42–43, 46–47
Pisa Customs House, 60
Pisano, Leonardo, 19, 22–23
popular math writing, 217–27
Porto Pisano, 56
practical arithmetic, 24
present value analysis, 208–9
printing press, 34
problem of the birds, 117–18,
    136–39

Ramanujan, Srinivasa, 69
recreational mathematics, 120, 133
Riccardiana Library, 170–80
Rinaldi, Gian Marco, 52–54,
    154–55, 163
Rule of three, 197–200

Scritti di Leonardo Pisano, 67
scuole d'abbaco, 26
Siena, 73–74
Siena manuscript of *Liber abbaci*,
    83, 140–45
Siena Public Library, 83, 139,
    141–45, 151–56

Sigler, Judith, 98–105
Sigler, Laurence, 39, 87, 94,
    98–101, 104–7
Silicon Valley, 33
Simon, Scott, 11
Sinus Pisanus, 56
Snow, C. P., 220
solidi, 62
State Archive of Pisa, 163–66

TeX, 110–114
trattati d'abbaco, 25
traveling merchant problems,
    204–5
Tucker, Anthony, 10

Vatican Library, 82, 139, 161–63
Vatican manuscript of *Liber abbaci*,
    82, 139, 161
Venice, 21, 56
Via Leonardo Fibonacci, 36, 149–50
Viète, Francois, 32
Vinci, Leonardo da, 47, 131, 155, 227

Weekend Edition, 11

Xuanpan, 31